Random Probability Measures on Polish Spaces

Stochastics Monographs

Theory and Applications of Stochastic Processes
*A series of books edited by Saul Jacka, Department of Statistics,
University of Warwick, UK*

Random Probability Measures on Polish Spaces

Hans Crauel

Department of Mathematics,
Technical University of Ilmenau,
Germany

CRC Press
Taylor & Francis Group
Boca Raton London New York

CRC Press is an imprint of the
Taylor & Francis Group, an **informa** business

CRC Press
Taylor & Francis Group
6000 Broken Sound Parkway NW, Suite 300
Boca Raton, FL 33487-2742

First issued in paperback 2019

© 2002 by Hans Crauel
CRC Press is an imprint of Taylor & Francis Group, an Informa business

No claim to original U.S. Government works

ISBN-13: 978-0-415-27387-9 (hbk)
ISBN-13: 978-0-367-39599-5 (pbk)

Visit the Taylor & Francis Web site at
http://www.taylorandfrancis.com

and the CRC Press Web site at
http://www.crcpress.com

Contents

Preface

The present work originates from joint work with Franco Flandoli on attractors for random dynamical systems, see [15]. The notion of a (global) attractor for a random dynamical system generalises that of an attractor for a deterministic system. As in the deterministic case, one is looking for a compact set attracting all orbits of the system. However, the attractor may depend on the random parameter. Furthermore, it is not supposed to stay fixed under the system. The system is allowed to move the (random) attractor, but only in a stationary manner, connected with the basic flow modelling the random influences.

In the deterministic case the notion of a global attractor has proved to be very useful for the study of certain infinite dimensional system, for instance for Navier–Stokes equations. For an actual account of the deterministic theory see, e.g., the book by Temam [36]. In order to obtain results for stochastic Navier–Stokes equations, the random dynamical systems taken under consideration were not confined to finite dimensional state spaces. They typically act on Banach or Hilbert spaces. This caused a problem: The methods available to obtain existence of invariant measures for random dynamical systems do not work in infinite dimensional spaces. This was the starting point of the present work.

The notion of a random dynamical system is a generalisation of a (deterministic) dynamical system. In the present work we are interested only in systems acting continuously on the state space. We have to distinguish between discrete and continuous time. In the case of discrete time a deterministic dynamical system is given by a continuous map φ from a topological space X to itself. For continuous time it is given by a (semi-) group $\varphi(t) : X \to X$, $t \in \mathbb{R}^+$ or $t \in \mathbb{R}$. We consider both the discrete and the continuous time case as (semi-) groups of continuous maps by taking the iterates $\varphi(n) = \varphi^n$, $n \in \mathbb{N}$ or $n \in \mathbb{Z}$, respectively, in the discrete time case. We denote by T the time, which is either $\mathbb{R}^+$, $\mathbb{R}$, $\mathbb{N}$, or $\mathbb{Z}$. The space X is called the state space of the system.

Roughly speaking, for a random dynamical system the map φ, or the (semi-) group $(\varphi(t))$, is allowed to be random. More precisely, a random dynamical system (RDS) is given by the following:

- time T, which is either $\mathbb{R}^+$, $\mathbb{R}$, $\mathbb{N}$, or $\mathbb{Z}$, and a probability space $(\Omega, \mathscr{F}, P)$ together with a measurable dynamical system $\vartheta_t : \Omega \to \Omega$, $t \in T$, such that P is invariant under ϑ_t for all $t \in T$

- a measurable map $\varphi : T \times X \times \Omega \to X$ such that $x \mapsto \varphi(t,\omega)x = \varphi(t,x,\omega)$ is continuous for all $t, \omega \in T \times \Omega$, and which satisfies the *cocycle property*, replacing the (semi-) group property: $\varphi(0,\omega) = \mathrm{id}|_X$, and

$$\varphi(t+s,\omega) = \varphi(t, \vartheta_s\omega) \circ \varphi(s,\omega)$$

for all $t, s \in T$ and all ω outside a P-nullset (which may depend on s, which gives a *crude* cocycle – see Definition 6.2 for complete details).

The concept of an RDS includes random and stochastic differential equations as well as products of random maps. In the context of stochastic differential equations it is closely related to *stochastic flows*. RDS have been dealt with under various aspects in the literature. For a thorough and comprehensive presentation see the monograph by Arnold [1], for some recent developments see the survey volume *Stochastic Dynamics* [16].

The concept of invariant measures is important for deterministic dynamical systems. It is considerably more important for RDS. This is due to the fact that other notions connected with long-term recurrent behaviour – such as fixed points and periodic orbits – occur considerably less for RDS than they do for deterministic systems. In order to study invariant measures for RDS we have to investigate the action of RDS on measures on the state space. This is not as straightforward as it is for deterministic systems. The classical approach goes back to Stochastic Analysis, where a generalisation of deterministic systems is used, which is quite different from the RDS approach. Stochastic Analysis does not substitute a deterministic flow $\varphi(t)$ by the stochastic flow $\varphi(t,\omega)$ (acting on the state space), but rather by a transition semigroup P_t, $t \geq 0$, acting on the space $C(X)$ of bounded continuous functions on the state space. The action of P_t is essentially given by mapping $f \in C(X)$ to $P_t f(x) = \int f(\varphi(t,\omega)x)\, dP(\omega)$. Then P_t is 'dualised', giving a semigroup acting on the space of Borel measures on the state space. Denoting this semigroup by P_t again, the action is essentially given by mapping a Borel measure ρ on X to $P_t\rho = \int \varphi(t,\omega)\rho\, dP(\omega) = E\varphi(t,\cdot)\rho$. So one obtains a semigroup on the space of measures on the state space.

The RDS approach, as opposed to the Stochastic Analysis approach, stays on the state space. The random map $\varphi(t, \omega) : X \to X$ maps a Borel measure ρ on X to $\varphi(t, \omega)\rho$, which is another Borel measure on X. The image measure $\varphi(t, \omega)\rho$ is, of course, random. Roughly speaking, one might say that the difference between the two approaches is that Stochastic Analysis takes expectation, whereas the RDS approach does not.

The image measure $\varphi(t, \omega)\rho$ being random, it is natural not to let $\varphi(t, \omega)$ act just on 'deterministic' measures on the state space, but to allow ρ to depend on ω, too. This leads to the question: What is a 'random measure' on X?

The investigation of random probability measures on Polish spaces is the major topic of the present work. Whereas it does make sense to consider the action of the RDS on general measures, probability measures are the most important class in the context of the theory of RDS. Here we will be concerned with probability measures only. Henceforth we always mean probability measures when speaking of measures unless indicated otherwise.

The book is organised as follows.

In Chapter 1 we introduce some basic notations and terminology, and provide some technical results on measurability and an estimate for integrals.

In Chapter 2 we introduce random sets and discuss their measurability properties. In particular, we introduce closed and open random sets, and give several equivalent characterisations. Then two major results on closed random sets are provided, which are the selection theorem and the projection theorem. The presentation given here follows Castaing and Valadier [9]. We finish Chapter 2 by proving that every compact random set is 'tight' in the sense that for any $\varepsilon > 0$ there exists a non-random compact set, containing the random one for all ω outside a set of P-measure not exceeding ε. Note that 'tight' here refers to an extension of the notion of tightness of a random variable, which means that the distribution of the random variable is concentrated on compact sets up to ε-mass. It does not mean tightness in the Polish space of compact sets equipped with the Hausdorff metric – a single compact random set is tight in this sense anyway.

In Chapter 3 we first introduce random probability measures as transition probabilities from Ω to X, identifying two of them if they coincide for P-almost all ω. Then we introduce the narrow topology on the space of random measures as the initial topology of the family of all *random* continuous bounded functions $f : X \times \Omega \to \mathbb{R}$. The 'random' narrow topology is thus defined in complete analogy with the usual narrow topology on the space of non-random probability measures on X, which is the initial topology of all (deterministic) continuous bounded functions $g : X \to \mathbb{R}$. We then show that the narrow topology is generated by random continuous functions taking values in the unit interval $[0, 1]$ already. Next, we obtain Theorem 3.17, which is an analogue of the Portmenteau theorem.

We continue by discussing the embedding of the space of deterministic probability measures into the random ones. The deterministic narrow topology coincides with the trace of the random narrow topology and with the quotient topology of the (continuous) map from the random to the deterministic measures obtained by mapping a random measure to its expectation.

Chapter 4 may be considered the main part of the book. It gives a generalisation of the Prohorov theorem for random measures. The classical Prohorov theorem states equivalence of tightness and relative compactness on the space of non-random measures. Following Valadier [37], we introduce the notion of tightness for random measures: A set of random measures is said to be tight if the set of non-random measures obtained by taking expectations is tight in the classical sense. We first get a characterisation of tightness in terms of compact random sets, which says that a set of random measures is tight if and only if the set is 'ω-wise tight in mean' (Proposition 4.3). In this form tightness comes up rather naturally in the context of RDS. For instance, the set of random measures supported by a random attractor can thus be shown to be tight. Theorem 4.4 says that also for random measures tightness and relative compactness (with respect to the narrow topology) are equivalent, and that both imply relative sequential compactness. It is almost immediate that relative compactness implies tightness (Lemma 4.5). To establish the other direction is more difficult. We first show that the narrow topology is induced already by evaluation in all random bounded Lipschitz functions, and in fact by all random Lipschitz functions such that both the function as well as the Lipschitz constant take values in the unit interval $[0, 1]$ (Corollary 4.10). With this approach we follow Dudley [19], Chapter 11.2–11.5, pp. 306–317. Note that the notion of a Lipschitz function refers to a metric on X, and the property of being Lipschitz as well as the Lipschitz constant depends on the choice of the metric. The main technical ingredients for the proof of the 'random Prohorov theorem' are Proposition 4.12 and Theorem 4.14. Proposition 4.12 uses the Stone–Daniell theorem to derive that every linear nonnegative functional on the space of random bounded Lipschitz functions with marginals P on Ω and a non-random Borel probability ρ on X is given by a random measure already. Then Theorem 4.14 states that the set of all random measures with marginal on X in a given compact set is compact itself. As a corollary we have that tightness implies relative compactness in the space of random measures. We then turn to the assertion that compactness implies sequential compactness in the space of random measures. In the non-random case this is automatic, since the space of non-random probability measures on a Polish space with the narrow topology is Polish itself. For the narrow topology on random measures this does not hold in general. In fact, the space of random measures with its narrow topology is Polish if and only if the σ-algebra of the underlying probability space is countably generated (mod P). We first show that the narrow topology on the space of random measures is metrisable in case of a countably generated (mod P) σ-algebra of the base space. This is achieved by constructing a metric, which is similar to a metric introduced by Dudley [19], Proposition 11.3.2, p. 310. We prove that this metric is also complete. Then we turn to conditional expectations in the space of random measures and show that the space of random measures which are measurable with respect to a sub-σ-algebra on the basic probability space is closed in the narrow topology. This allows to obtain sequential compactness from compactness by restriction to the σ-algebra (countably) generated by a sequence of random measures.

As one conclusion of Chapters 3 and 4 we obtain four characterisations of random measures on a Polish space X.

- By definition, random measures are transition probabilities from the probability space Ω to the state space X, identifying two of them if they coincide for P-almost all ω.

- Random measures are measures on the product space $X \times \Omega$ with marginal P (Proposition 3.3 (ii) and Proposition 3.6).

- Random measures are random variables taking values in the set $Pr(X)$ of non-random probability measures on X equipped with the Borel σ-algebra induced by the narrow topology on $Pr(X)$ (Remark 3.20 (i)).

- Random measures are nonnegative linear functionals on the linear space of random continuous functions on $X \times \Omega$ such that the marginal on X is a Borel probability measure and the marginal on Ω is P (Proposition 4.12). This means that evaluation of the functional in non-random continuous bounded functions gives the integral over a probability on X, and evaluation of the functional in functions depending only on ω gives the integral over P.

In Chapter 5 we introduce further topologies on the space of random measures by considering them as $Pr(X)$-valued random variables and integrating their distances. We obtain a list of topologies, the weakest of which is the topology of convergence in probability. We then show that the topology of convergence in probability is stronger than the narrow topology.

Thus the narrow topology is a fairly weak topology. In particular, the space of random measures over a compact space is always compact in the narrow topology, whereas in general the space of random measures is not compact in the topology of convergence in probability.

We proceed by considering an example which shows that convergence in the narrow topology of random measures does not imply convergence in law. Consequently, the map mapping a random measure to its law is not continuous with respect to narrow topology on the random measures and the narrow topology on the (deterministic) measures on the space of measures.

Chapter 6 introduces continuous RDS and investigates their action on random measures. A continuous RDS acts continuously on random measures when equipping them with the narrow topology. We obtain existence results for invariant measures provided the RDS leaves invariant a compact convex set of random measures. There are two methods to obtain existence of invariant measures. One is elegant and fast: Apply a fixed point theorem. The other one uses a Krylov–Bogolyubov argument, constructing an invariant measure as limit of time means. Existence of invariant measures can thus be established, in particular, as soon as there exists a compact random set which is invariant with respect to the cocycle. These invariant measures are supported by the compact random set.

Then we turn to RDS with two-sided base time. We introduce the past of the system to be the σ-algebra generated by the actions of the cocycle on the negative time axis. A random measure which is measurable with respect to the past of the system is said to be a *Markov measure*. Markov measures play a particular role for RDS induced by stochastic differential equations. They constitute the class of random measures which can be treated by means of Stochastic Analysis, see Crauel [11]. Assuming again existence of a compact random set which is invariant with respect to the cocycle and which in addition is measurable with respect to the past we obtain existence of invariant Markov measures supported by the compact random set. This is precisely the situation encountered when considering random attractors, see Crauel and Flandoli [15] and Crauel, Debussche and Flandoli [14].

We then turn to some ergodic theoretic considerations. For every random continuous function f we show that the infimum and the supremum of the integrals with respect to invariant measures are realised, and they are realised by ergodic invariant measures already.

In Appendix A we show that the narrow topology on the space of non-random probability measures on a Polish space is metrisable. This follows Dudley [19], Theorem 11.3.3, p. 310–311.

In Appendix B we collect some technical results.

In spite of the fact that many features of the Prohorov theory for non-random measures on a Polish space carry over to random measures, there is a major difference. The narrow topology on the space of Borel probability measures on a Polish space is separable, and it is metrisable by a complete metric. Thus the space of Borel probability measures on a Polish space with the narrow topology is Polish again. This does not carry over to random measures on a Polish space in general. The narrow topology on the space of random measures is neither metrisable nor separable in case the σ-algebra of the probability space is not countably generated (mod P). In particular, the space of random measures with the narrow topology is not Polish then. Thus it is not as self-evident as it is for non-random measures that tightness implies not only relative compactness, but also sequential relative compactness.

Random measures in the context of RDS have been dealt with by several authors, compare Arnold [1] for an introduction as well as for further references. Usually the state space X was assumed to be compact, or at least it was possible to find a compact subset of the state space invariant under the RDS. More generally, locally compact Hausdorff spaces with a countable base (LCCB) had been taken under consideration. If X is a compact metric space, or, more generally, if it is LCCB, then the space of random probability measures on X can be identified with a subset of $L^\infty(\Omega, \mathcal{M}(X))$, which is the space of P-essentially bounded $\mathcal{M}(X)$-valued functions, measurable with respect to the σ-algebra of the weak* topology on $\mathcal{M}(X)$. Here $C_0(X)$ denotes the space of continuous functions on X vanishing at infinity, and $\mathcal{M}(X)$ denotes the space of finite signed Borel measures on X. The space $L^\infty(\Omega, \mathcal{M}(X))$ can then be identified with the dual of $L^1(\Omega, C_0(X))$, invoking Bourbaki [8] Chapitre 6, § 2, N° 6. In this case the narrow topology on random probability measures can be identified as the (trace of the) weak* topology of the Banach space $L^\infty(\Omega, \mathcal{M}(X))$. See Arnold [1] 1.4–1.8, in particular 1.5, for details. One minor problem with this approach remains even for a compact metric X, insofar as Bourbaki [8] assumes $(\Omega, \mathcal{F}, P)$ to be a Borel measure space, and does not consider abstract probability spaces. Though this can be overcome, the approach still does not work if X is not LCCB.

One possibility to extend the approach described in the previous paragraph to Polish or to more general spaces consists in embedding the space X

into a compact metric space $\hat{X}$, see, e.g., Dellacherie and Meyer [18] or Williams [39]. The considerations of the previous paragraph can then be applied to the space of random measures on the bigger space $\hat{X}$. The hard part of the generalisation of the Prohorov theorem ('tight implies relatively compact' for probability measures on X) can be obtained by using the Portmenteau theorem (Theorem 3.17) on X, following Williams' proof of the classical Prohorov theorem (Williams [39], Theorem I.44, p. 25).

We have chosen not to follow this approach, but rather to stay on the given Polish space. In this respect we follow the presentation of the Prohorov theory as given by Dudley [19], 11.1–11.5, pp. 302–317. Instead of going into a big ambient space we rather imagine the space of random measures as a (measurable) bundle over the deterministic measures with bundle map given by the expectation. We then 'pull the classical Prohorov theorem up the fibres'. In particular, the present approach uses the classical Prohorov theorem in both directions (Lemma 4.5 and Corollary 4.15).

It should be emphasised that the narrow topology is custom designed for RDS. Its main use is to assure existence of invariant measures for RDS for which there is a compact random set which is invariant with respect to the RDS.

Random measures have been of interest in areas apart from RDS as well. Kallenberg [27] is concerned with random Radon measures on LCCB spaces, where his interest comes from point processes. A generalisation of the Prohorov theorem for relative sequential compactness in this situation can be found in [27] Lemma 4.5, p. 23.

The theory of super processes – which essentially are measure valued processes – also makes use of random measures. In fact, here random measures on more general spaces than Polish spaces become relevant, see, for example, Dawson and Perkins [17] or Etheridge [21]. In the field of super processes mainly the topology of convergence in probability on the space of random measures is of interest. The narrow topology on random measures has not been used. Recall that the narrow topology on random measures is weaker than the topology of convergence in probability.

Balder [3]–[5] and Valadier [37] use random measures in optimal control. Balder considers transition probabilities from a finite measure space to a standard Borel space (or, equivalently, metrisable Lusin space) in [3] and a σ-standard Borel space in [4]. He introduces a notion of tightness in terms of existence of inf-compact functions on the product space. Following Valadier [37], we refer to this notion as B-tightness. Balder then proves that B-tightness implies compactness and sequential compactness, see [3] Appendix A, pp. 591–596, and [4] Theorem 2.1, p. 431. An extension to transition probabilities from a complete σ-finite measure space to a completely regular Suslin space is given in Balder [5]. The methods used essentially proceed by embedding a standard Borel space into a compact metrisable space (Balder [3], Appendix A). The extensions in [4], [5] then build up on this result.

Valadier [37] is concerned with measures on the product space of a complete finite measure space with a metrisable Suslin space, whose marginal on the measure space is dominated by the given finite measure. Theorem 11 of [37], p. 162–163, states that tightness implies narrow relative compactness as well as narrow relative sequential compactness. The argument proceeds by embedding the Suslin space into a metrisable compact space.

As already pointed out, the present approach does not make use of an embedding of the state space into an ambient compact metric space. In this respect it differs from all other approaches to this problem (I am aware of). The question of metrisability of the narrow topology on the space of random measures, as well as the question for conditions under which the space of random measures equipped with the narrow topology is a Polish space, have not been considered previously.

We have tried to present the theory as elementarily as possible, avoiding deep measure theoretic results. The reader we have in mind is a student in the last third of his or her studies or a beginning graduate student with knowledge of a standard course in Real Analysis or 'Maß- und Wahrscheinlichkeitstheorie'. The only measure theoretic result which seems not to be treated in many standard courses is the Stone–Daniell theorem 4.11.[1]

In particular, we restrict attention to the simplest case, which is that of a Polish space. Several results can quite easily be extended to separable metric or more general spaces. In order not to overload the presentation we usually stick to Polish spaces, and do not care about the minimal assumptions under which a statement holds.

Another concern of the present approach is to avoid completion of the σ-algebra whenever possible. To achieve this we sometimes use 'brute force'. This means that whenever an argument produces a random variable measurable with respect to a completed σ-algebra only, we use the fact that for

---

[1]At least the Stone–Daniell theorem was not treated in full generality – but only in the restricted version given by the Riesz representation theorem – in the courses I became aware of.

random variables measurable with respect to the completion one can find another random variable measurable with respect to the original σ-algebra such that the two coincide outside a set of measure zero. This holds for random variables taking values in a separable metric space (compare Lemma 1.2), for 'random continuous functions' (Lemma 1.3), and for 'closed random sets' (Lemma 2.7).

Example 5.7 goes back to an idea of Gero Fendler, Saarbrücken. I am grateful to him and to Vibeke Thinggaard, Kopenhagen, for many inspiring discussions. I am also indebted to Ludwig Arnold, Bremen, for his support over many years, and to Franco Flandoli, Pisa, who encouraged me to leave compact metric spaces.

Chapter 1

Notations and Some Technical Results

In this brief chapter we first introduce some basic notations. We refer to standard textbooks on Real Analysis or on measure and integration theory for the notions used.

Then technical results on measurability and completion as well as on approximation of integrals are provided.

Notations

Suppose that $(Y, \mathcal{Y})$ is a measurable space, so Y is a set, and $\mathcal{Y}$ is a σ-algebra of subsets of Y. For $(Z, \mathcal{Z})$ another measurable space a map $f : Y \mapsto Z$ is measurable if $f^{-1}(B) \in \mathcal{Y}$ for every $B \in \mathcal{Z}$.

Suppose that $(Y, \mathcal{Y}, m)$ is a measure space, so $m : \mathcal{Y} \to [0, \infty]$ is a σ-additive set function. The image measure of m under a measurable $f : Y \to Z$ is denoted by fm, $fm(B) = m(f^{-1}B)$ for $B \in \mathcal{Y}$. In the following we will be concerned with probability measures only. The set of probability measures over $(Y, \mathcal{Y})$ will be denoted by $Pr(Y, \mathcal{Y})$ or by $Pr(Y)$, if there is no ambiguity what the σ-algebra $\mathcal{Y}$ is concerned.

Suppose that $(X, \mathcal{T})$ is a topological space, so X is a set, and $\mathcal{T}$ is the system of open subsets of X. Then X is considered as a measurable space with its Borel σ-algebra, which is the smallest σ-algebra containing all open sets. The Borel σ-algebra is denoted by $\mathcal{B}(X)$ or just $\mathcal{B}$.

1

A Polish space X is a separable topological space whose topology is metrisable by a complete metric. A metric on a Polish space is always understood to be complete in the following.

If X is a metric space we denote by $B(x, \delta)$ or by $B_\delta(x)$ the open ball of radius δ around $x \in X$.

For a set Y and subsets $A, B \subset Y$ we denote by $A^c = Y \backslash A = \{y \in Y : y \notin A\}$ the complement of A in Y, and by $A \triangle B = (A \backslash B) \cup (B \backslash A) = (A \cap B^c) \cup (B \cap A^c)$ the symmetric difference of A and B. If $(\Omega, \mathscr{F}, P)$ is a probability space, the σ-algebra $\mathscr{F}^0 = \{F \triangle M : M \subset \mathscr{F}$ with $M \subset N$ for some $N \in \mathscr{F}$ with $P(N) = 0\}$ is said to be the *completion of* $\mathscr{F}$ (with respect to P).

Measurability and Completion

We now collect some technical results on measurability and completion.

The assertion of the next lemma can be found in several places, see, e.g., Kuratowski ([29] p. 378), or Himmelberg ([26], Theorem 6.1, pp. 64–65).

1.1 Lemma *Suppose that* Y *is a separable metric space, that* $(\Omega, \mathscr{F})$ *is a measurable space, and let* Z *be another metric space. Suppose that* $f : Y \times \Omega \to Z$ *satisfies*

   (i) $\omega \mapsto f(y, \omega)$ *is measurable for each* $y \in Y$

   (ii) $y \mapsto f(y, \omega)$ *is continuous for each* $\omega \in \Omega$

Then f *is jointly measurable.*

PROOF Let D be a countable dense set in Y. Given $n \in \mathbb{N}$, choose a subset $D_n \subset D$ and a partition $\mathfrak{U}_n$ of Y, consisting of measurable subsets, such that $\mathfrak{U}_n(y) \subset B(y, 1/n)$ for all $y \in D_n$, where $\mathfrak{U}(y)$ denotes the element of the partition $\mathfrak{U}$ containing y.

Put $f_n(x, \omega) = \sum_{y \in D_n} f(y, \omega) 1_{\mathfrak{U}_n(y)}(x)$ (where the sum is to be understood as an abbreviating notation). If $G \subset Z$ is a Borel set, then

$$f_n^{-1}(G) = \bigcup_{y \in D_n} \mathfrak{U}_n(y) \times f(y, \cdot)^{-1}(G),$$

hence f_n is jointly measurable. On the other hand, for $(x, \omega) \in Y \times \Omega$,

$$d(f(x, \omega), f_n(x, \omega)) \leq \sup_{y : d(x, y) < 1/n} d(f(x, \omega), f(y, \omega)),$$

hence $f(x, \omega) = \lim_{n \to \infty} f_n(x, \omega)$ for all $(x, \omega) \in Y \times \Omega$. Consequently, f is jointly measurable. $\qquad \square$

1.2 Lemma *Suppose that $(\Omega, \mathscr{F}, P)$ is a probability space, and let $\mathscr{F}^0$ be the completion of $\mathscr{F}$ with respect to P. Let Y be a separable metric space. Then for any $\mathscr{F}^0$-measurable map $f^0 : \Omega \to Y$ there exists an $\mathscr{F}$-measurable map $f : \Omega \to Y$ with $f = f^0$ almost surely with respect to (the completion of) P.*

PROOF As in the proof of Lemma 1.1, let $D \subset Y$ be countable and dense in Y. For $n \in \mathbb{N}$ choose $D_n = \{y_n^k : k \in \mathbb{N}\} \subset D$ and a partition $\mathfrak{B}_n$ into measurable subsets of Y such that $B_n^k \subset B(y_n^k, 1/n)$ for all $k \in \mathbb{N}$, where $B_n^k = \mathfrak{B}_n(y_n^k)$ is the element of $\mathfrak{B}_n$ containing y_n^k. (One could choose $\mathfrak{B}_n$ increasing with n, but it is not necessary to do so at this point.) Define a sequence $(f_n^0)_{n \in \mathbb{N}}$ of $\mathscr{F}^0$-measurable maps by

$$f_n^0(\omega) = \sum_{k \in \mathbb{N}} y_n^k 1_{U_n^k}(\omega),$$

where $U_n^k = \{\omega : f^0(\omega) \in B_n^k\}$ (again the sum is only an abbreviating notation). Then

$$d(f_n^0(\omega), f^0(\omega)) \leq \frac{1}{n}$$

for all $\omega \in \Omega$; hence f_n^0 converges to f^0 P-almost surely (P-a.s.) (in fact, even uniformly). Since $U_n^k \in \mathscr{F}^0$ there exist $F_n^k \in \mathscr{F}$ with $F_n^k \subset U_n^k$ and $P(U_n^k \setminus F_n^k) = 0$. For each $n \in \mathbb{N}$ put $F_n^\infty = \left(\bigcup_{k \in \mathbb{N}} F_n^k \right)^c$; then $\mathfrak{F}_n = \{F_n^k : 0 \leq k \leq \infty\}$ is a partition of Ω into $\mathscr{F}$-measurable sets. Put

$$f_n(\omega) = \sum_{0 \leq k \leq \infty} y_n^k 1_{F_n^k}(\omega),$$

where $y_n^\infty \in Y$ is arbitrary. Then P-a.s. $f_n = f_n^0$. Since f_n^0 converges P-a.s. for $n \to \infty$, also f_n converges P-a.s. Put $f = \lim f_n$, then f is $\mathscr{F}$-measurable, and P-a.s. $f = \lim f_n = \lim f_n^0 = f^0$. $\qquad \square$

In the following lemma we encounter the problem that not every continuous map from a non-closed subset of a separable metric space X to another separable metric space Y can be extended to a continuous map on the whole space. Consider, e.g., the function $x \mapsto \sin(1/x)$ on the interval $(0, 1]$, which cannot be extended to be continuous at 0. But note that if $D \subset X$ is a dense subset of X then for any two continuous functions $g : X \to Y$ and $h : X \to Y$ we have $g = h$ if and only if $g(x) = h(x)$ for all $x \in D$. Hence a continuous function $f : D \to Y$ possibly cannot be extended. But as soon as it can be extended, the extension is unique. Furthermore, the extension is determined by the values $\{f(x) : x \in D\}$ ($f(y)$ is the limit of $f(x_n)$ for every sequence x_n, $n \in \mathbb{N}$, with $x_n \in D$ converging to y).

1.3 Lemma *Suppose that X and Y are separable metric spaces, $(\Omega, \mathscr{F}, P)$ is a probability space, $\mathscr{F}^0$ is the completion of $\mathscr{F}$ with respect to P, and $f^0 : X \times \Omega \mapsto Y$ satisfies*

(i) $\omega \mapsto f^0(x,\omega)$ *is measurable with respect to $\mathscr{F}^0$ for each $x \in X$,*

(ii) $x \mapsto f^0(x,\omega)$ *is continuous for each $\omega \in \Omega$.*

Then there exists $f : X \times \Omega \to Y$, such that

(i) $\omega \mapsto f(x,\omega)$ *is measurable with respect to $\mathscr{F}$ for each $x \in X$,*

(ii) $x \mapsto f(x,\omega)$ *is continuous for each $\omega \in \Omega$,*

and $f(\cdot,\omega) = f^0(\cdot,\omega)$ for P-almost all $\omega \in \Omega$.

PROOF Let $D \subset X$ be countable and dense in X. For every $d \in D$ Lemma 1.2 yields existence of an $\mathscr{F}$-measurable $F_d : \Omega \to Y$ with $F_d(\omega) = f^0(d,\omega)$ for P-almost all ω. By countability of D we get $F_d(\omega) = f^0(d,\omega)$ for all $d \in D$ and for all ω outside a nullset $N_0 \in \mathscr{F}^0$. For $\omega \in N_0^c$ the map $d \mapsto F_d(\omega)$ is the restriction of the continuous map $d \mapsto f^0(d,\omega)$. Choose $N \in \mathscr{F}$ with $N_0 \subset N$ and $P(N) = 0$. For $\omega \in N^c$ we thus have

$$\lim_{n \to \infty} \inf_{d \in B(x,1/n) \cap D} F_d(\omega) = \lim_{n \to \infty} \sup_{d \in B(x,1/n) \cap D} F_d(\omega) = f^0(x,\omega)$$

for all $x \in X$. Let $g : X \to Y$ be an arbitrary continuous map, e.g., a constant, and define

$$f(x,\omega) = \begin{cases} f^0(x,\omega) & \text{for } \omega \in N^c, \\ g(x) & \text{for } \omega \in N. \end{cases}$$

Then, for every $x \in X$, $\omega \mapsto f(x,\omega)$ is measurable with respect to $\mathscr{F}$, since for any Borel set $U \subset Y$

$$\{\omega : f(x,\omega) \in U\}$$
$$= \left(\{\omega : \lim_{n \to \infty} \inf_{d \in B(x,1/n) \cap D} F_d(\omega) \in U\} \cap N^c\right) \cup \left(\{\omega : g(x) \in U\} \cap N\right).$$

Furthermore, $f(\cdot,\omega) = f^0(\cdot,\omega)$ for every $\omega \in N^c$, and $P(N^c) = 1$. $\square$

Finally, we provide a lemma relating the integral of a function with the integrals of its approximations by piecewise constant functions. Since this relation will be used in later chapters at several places, we give the details of the proof.

1.4 Lemma *Let $(X, \mathscr{B})$ be a measurable space, and suppose that $f : X \to [0,1]$ is measurable. For $n \in \mathbb{N}$ put $B_k = \{x : f(x) \geq \frac{k}{n}\}$, $0 \leq k \leq n+1$ (so $B_0 = X$ and $B_{n+1} = \emptyset$). Then, for any finite measure ν on $(X, \mathscr{B})$,*

$$\frac{1}{n}\sum_{k=1}^{n}\nu(B_k) \leq \int f \, d\nu \leq \frac{1}{n}\sum_{k=0}^{n}\nu(B_k).$$

PROOF We have

$$\sum_{k=0}^{n}\frac{k}{n}\nu(B_k \setminus B_{k+1}) \leq \int f \, d\nu \leq \sum_{k=0}^{n}\frac{k+1}{n}\nu(B_k \setminus B_{k+1}),$$

hence, invoking $\nu(B_k \setminus B_{k+1}) = \nu(B_k) - \nu(B_{k+1})$,

$$\frac{1}{n}\sum_{k=0}^{n}k\big(\nu(B_k) - \nu(B_{k+1})\big) \leq \int f \, d\nu \leq \frac{1}{n}\sum_{k=0}^{n}(k+1)\big(\nu(B_k) - \nu(B_{k+1})\big).$$

Now

$$\begin{aligned}
\sum_{k=0}^{n}k\big(\nu(B_k) - \nu(B_{k+1})\big) &= \sum_{k=0}^{n}\big(k\nu(B_k) - (k+1)\nu(B_{k+1})\big) + \sum_{k=0}^{n}\nu(B_{k+1}) \\
&= \sum_{k=0}^{n}\nu(B_{k+1}) = \sum_{k=1}^{n}\nu(B_k),
\end{aligned}$$

hence

$$\begin{aligned}
\sum_{k=0}^{n}(k+1)\big(\nu(B_k) - \nu(B_{k+1})\big) &= \sum_{k=1}^{n}\nu(B_k) + \sum_{k=0}^{n}\nu(B_k) - \nu(B_{k+1}) \\
&= \sum_{k=1}^{n}\nu(B_k) + \nu(B_0) \\
&= \sum_{k=0}^{n}\nu(B_k).
\end{aligned}$$

The assertion follows. $\square$

Chapter 2

Random Sets

This chapter discusses closed and open random sets. Firstly, the definition and several equivalent characterisations are given. Then we turn to the two major results on closed random sets, which are the selection theorem and the projection theorem. The presentation given here follows Castaing and Valadier [9]. Finally, we obtain a proposition which says that a compact random set in a Polish space is 'tight' in the sense that the random set is subset of a compact deterministic set with probability arbitrarily close to 1. This property is of interest for random attractors. It is used, e.g., to prove the uniqueness of global set attractors, see Crauel [13].

Suppose that $(\Omega, \mathscr{F}, P)$ is a probability space. We will be concerned with set valued maps $A : \Omega \to 2^X$, where 2^X denotes the set of all subsets of X. A set valued map A is uniquely determined by its graph

$$\operatorname{graph}(A) = \{(x, \omega) : x \in A(\omega)\}$$

which is a subset of $X \times \Omega$. For a subset Γ of $X \times \Omega$ we denote by $\Gamma(\omega) = \Gamma_\omega = \{x \in X : (x, \omega) \in \Gamma\}$ the ω-sections of Γ, thus defining a set valued map (possibly empty-valued). For $A \subset X$ denote by $A^c = X \setminus A$ the complement of A in X. For a set valued map A note that $\operatorname{graph}(A^c) = \big(\operatorname{graph}(A)\big)^c$. Most of the present chapter refers only to the measurable space $(\Omega, \mathscr{F})$. It is only Lemma 2.7 and Proposition 2.15 which refer to the particular probability P.

We need some more notations. If X is a topological space, then $\overline{A} = \operatorname{cl}(A)$ denotes the closure of A in X, and $\operatorname{int}(A)$ denotes the interior of A. For a metric space (X, d) and for $x \in X$, $A \subset X$ put $d(x, A) = \inf\{d(x, a) : a \in A\}$. Continuity of $a \mapsto d(x, a)$ implies

$$d(x, A) = d(x, \overline{A}) \tag{2.1}$$

for any $A \subset X$ and $x \in X$.

For $A \subset X$ and $\delta > 0$ denote by A^δ the δ-neighbourhood of A,

$$
\begin{aligned}
A^\delta &= \{x \in X : d(x, A) < \delta\} = \{x : d(x, a) < \delta \text{ for some } a \in A\} \\
&= \{x : B_\delta(x) \cap A \neq \emptyset\}
\end{aligned}
$$

where $B_\delta(x)$ is the open ball of radius δ around x. It is immediate from (2.1) that $(\overline{A})^\delta = A^\delta$ for $A \subset X$ arbitrary.

Closed and Open Random Sets

2.1 Definition (CLOSED AND OPEN RANDOM SETS) A set valued map $C : \Omega \to 2^X$ taking values in the closed subsets of a Polish space X is said to be *measurable* if for each $x \in X$ the map $\omega \mapsto d(x, C(\omega))$ is measurable. In this case C is called a *closed random set.*
A set valued map $\omega \mapsto U(\omega)$ is said to be an *open random set* if its complement U^c is a closed random set.

Random sets have been investigated by Castaing and Valadier [9], who addressed them mainly as *measurable multifunctions*. The notion of closed and open random sets will be discussed after Proposition 2.4.

For an arbitrary (not necessarily open or closed) set valued mapping $\omega \mapsto A(\omega)$, measurability of $\omega \mapsto d(x, A(\omega))$ for each $x \in X$ is equivalent to $\omega \mapsto \overline{A}(\omega)$ being a closed random set. This is immediate from (2.1) again.

2.2 Remark Suppose that $\omega \mapsto A(\omega)$ is an arbitrary (not necessarily open or closed) set valued map such that $\omega \mapsto d(x, A(\omega))$ is measurable for every $x \in X$. Since $x \mapsto d(x, A(\omega))$ is continuous for every ω, measurability of $(x, \omega) \mapsto d(x, A(\omega))$ follows from Lemma 1.1.

2.3 Definition Let $(\Omega, \mathscr{F})$ be a measurable space. The *σ-algebra of universally measurable sets associated with $\mathscr{F}$*, or *universal completion of $\mathscr{F}$*, is $\bigcap_Q \mathscr{F}_Q$, where $\mathscr{F}_Q$ denotes the completion of $\mathscr{F}$ with respect to a positive measure Q on $(\Omega, \mathscr{F})$, and the intersection is taken over all positive finite measures Q. We will speak of the *universally completed σ-algebra of $\mathscr{F}$*, and call its elements *universally measurable sets.*

Any σ-algebra $\mathscr{F}$ containing its universally completed σ-algebra already coincides with it. In this case we say that $\mathscr{F}$ is *universally complete* or just *universal.*

Note that any σ-algebra which is complete with respect to some finite measure contains its universal σ-algebra, so it must coincide with its universal

σ-algebra. Therefore the σ-algebra of a complete probability space is automatically universally complete.

The following Proposition has considerable overlap with Theorem III.30 of Castaing and Valadier ([9], p. 80).

2.4 Proposition (CHARACTERISATIONS OF CLOSED RANDOM SETS) *For a set valued map $C : \Omega \to 2^X$, taking values in the closed subsets of a Polish space X, consider the following conditions:*

(i) *$\omega \mapsto d(x, C(\omega))$ is measurable for every $x \in X$ (i.e., C is a closed random set)*

(ii) *for all open $U \subset X$ the set $\{\omega : C(\omega) \cap U \neq \emptyset\}$ is measurable*

(iii) *for every $\delta > 0$, graph(C^δ) is a measurable subset of $X \times \Omega$, where graph(C^δ) is the graph of the δ-neighbourhood $\omega \mapsto C^\delta(\omega)$ of C*

(iv) *graph(C) is a measurable subset of $X \times \Omega$*

Then (i), (ii) and (iii) are equivalent, and either of them implies (iv). Furthermore, (iv) implies (i) if $\mathscr{F}$ is universally complete.

PROOF In order to see that (i) implies (ii) let $U \subset X$ be open, and let $D \subset X$ be dense and separable. For every $x \in D \cap U$ choose α_x with $d(x, U^c)/2 < \alpha_x < d(x, U^c)$, say. Then $B(x, \alpha_x) \subset U$ and $U = \bigcup_{x \in D \cap U} B(x, \alpha_x)$. Thus $\{\omega : C(\omega) \cap U \neq \emptyset\} = \bigcup_{x \in D}\{\omega : C(\omega) \cap B(x, \alpha_x) \neq \emptyset\} = \bigcup_{x \in D}\{\omega : d(x, C(\omega)) < \alpha_x\}$, which is measurable.

To show that (ii) implies (i) let $x \in X$ and $\delta > 0$. Then $\{\omega : d(x, C(\omega)) < \delta\} = \{\omega : C(\omega) \cap B(x, \delta) \neq \emptyset\} \in \mathscr{F}$ by (ii) applied with $U = B(x, \delta)$. This being true for every $\delta > 0$, $\omega \mapsto d(x, C(\omega))$ is measurable, which is (i).

To prove the equivalence of (i) and (iii), first suppose that (i) holds. Then

$$\text{graph}(C^\delta) = \{(x, \omega) : x \in C^\delta(\omega)\} = \{(x, \omega) : d(x, C(\omega)) < \delta\},$$

hence measurability of graph(C^δ) follows from joint measurability of $(x, \omega) \mapsto d(x, C(\omega))$, see Remark 2.2.

If on the other hand graph(C^δ) is a measurable subset of $X \times \Omega$ for every $\delta > 0$, then for every $x \in X$ it holds that $\{\omega : d(x, C(\omega)) < \delta\} \in \mathscr{F}$ for all $\delta > 0$. But this means measurability of $\omega \mapsto d(x, C(\omega))$, which is (i).

To see that (i) implies (iv) note that graph$(C) = \{(x, \omega) : x \in C(\omega)\} = \{(x, \omega) : d(x, C(\omega)) = 0\}$.

For the proof that (iv) implies (i) in case $\mathscr{F}$ is universally complete we need the Projection Theorem (Theorem 2.12). $\qquad\square$

2.5 Remark (i) Condition (ii) of Proposition 2.4 refers only to the topology of X, not to the metric. Consequently, measurability of closed set valued maps does not depend on the choice of the metric d, as might be suggested by Definition 2.1.

(ii) In Proposition 2.4 (iii) it suffices to have measurability of graph(C^δ) for every δ from a dense subset of $\mathbb{R}^+$, e.g., for $\delta \in \mathbb{Q}^+$.

(iii) If A is an arbitrary set valued map, then $\overline{A}^\delta = A^\delta$, and hence Proposition 2.4 implies that graph(A^δ) is measurable for all $\delta > 0$ if and only if $\overline{A}$ is a closed random set.

(iv) For an arbitrary set valued $\omega \mapsto A(\omega)$ condition (iv) of Proposition 2.4 is a much more distinguishing property than (i), (ii) and (iii), which all only refer to properties of $\omega \mapsto \overline{A}(\omega)$. In fact, for $A \subset X$ arbitrary and $U \subset X$ open $A \cap U = \emptyset \Rightarrow A \subset U^c \Rightarrow \overline{A} \subset U^c \Rightarrow U \cap \overline{A} = \emptyset \Rightarrow U \cap A = \emptyset$, using U^c closed.

An alternative notion of a closed random set $\omega \mapsto C(\omega)$ would be: Take a set $C \subset \mathscr{B} \otimes \mathscr{F}$ such that all (or almost all) of its sections $\omega \mapsto C(\omega)$ are closed, and call this a closed random set. Proposition 2.4 then gives measurability of $\omega \mapsto d(x, C(\omega))$, $x \in X$, only with respect to the universal σ-algebra of $\mathscr{F}$. Since we want to use measurability with respect to the original $\mathscr{F}$, we chose the stronger notions of closed and open random sets, respectively, as introduced in Definition 2.1. We will sometimes address the sections $\omega \mapsto A(\omega)$ of $A \subset \mathscr{B} \otimes \mathscr{F}$ as a random set. This causes a slight inconsistency – a random set which is closed for all ω is not necessarily a closed random set. Lemma 2.7 will show that this is not so bad: For any random set which is closed for all ω there exists a closed random set such that the two sets coincide for P-almost all ω.

The Selection Theorem

2.6 Theorem (SELECTION THEOREM) *A set valued map $C : \Omega \to 2^X$, taking values in the closed nonvoid subsets of a Polish space X, is a closed random set if, and only if, there exists a sequence $(c_n)_{n\in\mathbb{N}}$ of measurable maps $c_n : \Omega \to X$, such that*

$$C(\omega) = \mathrm{cl}\left\{c_n(\omega) : n \in \mathbb{N}\right\}$$

for all $\omega \in \Omega$ (cl denoting closure). In particular, if C is a closed random set *then there exists a measurable selection, i.e., a measurable map $c : \Omega \to X$ such that $c(\omega) \in C(\omega)$ for all $\omega \in \Omega$.*

For the proof see Castaing and Valadier [9], Theorem III.9, p. 67.

2.7 Lemma *Suppose that $C^0 : \Omega \to 2^X$ is a set valued map taking values in the closed subsets of a Polish space X such that $\mathrm{graph}(C)$ is a measurable subset of $\mathscr{B} \otimes \mathscr{F}^0$, where $\mathscr{F}^0$ denotes the completion of $\mathscr{F}$ with respect to P. Then there exists a closed random set $C : \Omega \to 2^X$ (with respect to $\mathscr{F}$) with $C(\omega) = C^0(\omega)$ for P-almost all $\omega \in \Omega$.*

PROOF   First suppose C^0 is nonvoid for all $\omega \in \Omega$. Then by Theorem 2.6 there exists a sequence $(c_n^0)_{n \in \mathbb{N}}$ of $\mathscr{F}^0$-measurable maps $c_n^0 : \Omega \to X$ such that $C^0(\omega) = \mathrm{cl}\{c_n^0(\omega) : n \in \mathbb{N}\}$. For every c_n^0 there exists $c_n : \Omega \to X$, measurable with respect to the uncompleted $\mathscr{F}$, such that $c_n = c_n^0$ P-almost surely by Lemma 1.2. Put $C(\omega) = \mathrm{cl}\{c_n(\omega) : n \in \mathbb{N}\}$. Then C is measurable with respect to $\mathscr{F}$ again by Theorem 2.6, and $\{\omega : C(\omega) \neq C^0(\omega)\}$ is a P-nullset.

For C^0 not necessarily nonvoid let $F_0 = \{\omega : C^0(\omega) = \emptyset\} \in \mathscr{F}^0$, and choose $F \in \mathscr{F}$ with $P(F \,\triangle\, F_0) = 0$. Put

$$\hat{C}^0(\omega) = \begin{cases} C^0(\omega) & \text{for } \omega \in F_0^c, \\ X & \text{for } \omega \in F_0. \end{cases}$$

Then $\hat{C}^0$ is a nonvoid closed random set with respect to $\mathscr{F}^0$. Consequently, there exists a closed random set $\omega \mapsto \hat{C}(\omega)$ with respect to $\mathscr{F}$ such that $\hat{C} = \hat{C}^0$ P-a.s. Put

$$C(\omega) = \begin{cases} \hat{C}(\omega) & \text{for } \omega \in F^c, \\ \emptyset & \text{for } \omega \in F. \end{cases}$$

Then C is $\mathscr{F}$-measurable, and $C = C^0$ P-a.s. $\square$

2.8 Remark (i) If C_1 and C_2 are closed random sets in a Polish space then $\{\omega : C_1(\omega) \subset C_2(\omega)\}$ and $\{\omega : C_1(\omega) = C_2(\omega)\}$ are measurable. This is immediate from Definition 2.1, since $C_1 \subset C_2$ if and only if $d(x, C_1) \geq d(x, C_2)$ for all x from a dense set in X.

(ii) If $A : \Omega \to 2^X$ satisfies $\mathrm{graph}(A) \subset \mathscr{B} \otimes \mathscr{F}$, then $(x, \omega) \mapsto 1_{A(\omega)}(x)$ is the indicator of $\mathrm{graph}(A)$, and hence it is jointly measurable. This applies, in particular, in the case where A is a closed or an open random set, using Proposition 2.4.

2.9 Proposition *If $\omega \mapsto C(\omega)$ is a closed random set in a separable metric space X then $\overline{C^c}$ is a closed random set as well.*

PROOF By Proposition 2.4 (ii) it suffices to prove $\{\omega : \overline{C^c}(\omega) \cap U = \emptyset\} \in \mathscr{F}$ for every open $U \subset X$. Now, for U open and A arbitrary, $U \cap A = \emptyset$ if and only if $U \cap \overline{A} = \emptyset$ (see Remark 2.5 (iv)). Thus $\overline{C^c} \cap U = \emptyset$ if and only if $C^c \cap U = \emptyset$, which is equivalent to $U \subset C$. Since C is assumed to be closed, $U \subset C$ if and only if $d(x, C) = 0$ for every $x \in U$, and this holds if and only if $d(x, C) = 0$ for every x from a dense subset of U. So let D be a countable dense subset of U (which exists by separability of X). Then

$$\{\omega : \overline{C^c}(\omega) \cap U = \emptyset\} = \{\omega : U \subset C(\omega)\} = \bigcap_{x \in D} \{\omega : d(x, C(\omega)) = 0\}$$

which implies measurability of $\omega \mapsto \overline{C^c}(\omega)$. $\square$

2.10 Corollary *Suppose that $A : \Omega \to 2^X$ is a set valued map on a Polish space X. Then*

  (i) *If A is a closed random set then $\omega \mapsto d(x, A^c(\omega))$ is measurable for each $x \in X$.*

 (ii) *If A is an open random set then $\omega \mapsto \overline{A}(\omega)$ is a closed random set. In particular, $\omega \mapsto d(x, A(\omega))$ is measurable for each $x \in X$.*

(iii) *If A is a closed random set then its interior $\operatorname{int} A$ is an open random set.*

PROOF From (2.1) we get $d(x, A^c) = d(x, \overline{A^c})$ and $d(x, A) = d(x, \overline{A})$, and hence both (i) and (ii) are immediate from Proposition 2.9. To establish (iii) we have to prove that $(\operatorname{int} A)^c$ is a closed random set. Since $(\operatorname{int} E)^c = \overline{E^c}$ for an arbitrary subset E of an arbitrary topological space, $(\operatorname{int} A)^c = \overline{A^c}$ is a closed random set by Proposition 2.9. $\square$

The next remark discusses measurability questions for set valued maps. It will not be used in the following.

2.11 Remark (i) For an open set valued $\omega \mapsto U(\omega)$ it does not suffice to know that $\overline{U}$ is a closed random set to conclude that U is an open random set. For instance, if $U(\omega) = X \setminus \{x_0\}$ for ω from a non-measurable set and $U(\omega) = X$ otherwise, then $\overline{U}$ is measurable, and U is not. Thus, U being an open random set implies that $d(x, U(\cdot))$ is measurable for every $x \in X$, but not vice versa.

(ii) Suppose that C is a closed random set. It seems natural to ask whether this implies that C^δ is an open random set for $\delta > 0$. As soon as the σ-algebra $\mathscr{F}$ is universal then this holds true by Proposition 2.4. It seems that C^δ need not be an open random set in general without $\mathscr{F}$ being universally completed. As soon as the closure of C^δ is a closed random set then C^δ is an open random set by Corollary 2.10 (iii). But note that $\overline{C^\delta} \neq \{x : d(x, C(\omega)) \leq \delta\}$ in general. This is a problem already for the case where C is a one point set $\omega \mapsto \{x(\omega)\}$, induced by a random variable x. Then C^δ is an open δ-ball with random center x. This question depends on the choice of the metric d.

(iii) Suppose that $\mathscr{F}$ is not universal. Suppose further that C is a set valued map which takes values in the closed sets, and which has a measurable graph, but which is not a closed random set (so $d(x, C(\omega))$ is not measurable for some x, or $\{\omega : C(\omega) \cap U \neq \emptyset\} \notin \mathscr{F}$ for some open $U \subset X$). Then the set of $\delta > 0$ such that $\mathrm{graph}(C^\delta) \notin \mathscr{B} \otimes \mathscr{F}$ cannot be the complement of a dense subset of $[0, \infty)$ by Remark 2.5 (ii). Therefore it must contain an open interval (which will depend on the choice of d). This shows the relation between condition (iv) on the one hand and (the equivalent) conditions (i), (ii) and (iii) of Proposition 2.4 on the other hand. Note that this only occurs in case $\mathscr{F}$ is not universally complete, and, in particular, not complete with respect to some probability measure. Otherwise all four conditions of Proposition 2.4 are equivalent.

The Projection Theorem

We further need the projection theorem of Castaing and Valadier [9] (Theorem III.23, p. 75).

2.12 Theorem (PROJECTION THEOREM) *If $(\Omega, \mathscr{F})$ is a measurable space, and if X is a Polish space, then the projection of $A \subset X \times \Omega$ to Ω, given by*

$$\pi_\Omega(A) = \{\omega : (x, \omega) \in A \text{ for some } x \in X\},$$

is universally measurable for any $A \in \mathscr{B} \otimes \mathscr{F}$ (thus, in particular, $\pi_\Omega(A)$ is an element of the completion of $\mathscr{F}$ with respect to any probability measure on $(\Omega, \mathscr{F})$).

As an immediate consequence of the projection theorem we get that taking suprema or infima, respectively, over an arbitrary family of functions gives a universally measurable function, provided the family of functions depends on the argument *and* the parameter in a jointly measurable way. First consider a more general formulation.

2.13 Corollary *Suppose that $f : X \times \Omega \to \mathbb{R}$ is measurable, where $(\Omega, \mathscr{F})$ is a measurable space and X is a Polish space. Let $\omega \mapsto C(\omega)$ be any set valued mapping such that $\mathrm{graph}(C)$ is measurable. Then*

$$\omega \mapsto \sup_{x \in C(\omega)} f(x, \omega)$$

is measurable with respect to the universally completed σ-algebra of $\mathscr{F}$.

PROOF For any $\alpha \in \mathbb{R}$

$$\left\{\omega : \sup_{x \in C(\omega)} f(x, \omega) > \alpha\right\} = \pi_\Omega\Big[\{(x, \omega) : f(x, \omega) > \alpha\} \cap \mathrm{graph}(C)\Big]$$

is measurable with respect to the universally completed σ-algebra of $\mathscr{F}$ by Theorem 2.12. $\square$

Corollary 2.13 applies, in particular, to $C(\omega) = X$, so for any measurable $f : X \times \Omega \to \mathbb{R}$

$$\omega \mapsto \sup_{x \in X} f(x, \omega)$$

is universally measurable.

REMAINDER OF THE PROOF OF PROPOSITION 2.4 We have to prove that any closed set valued $C : \Omega \to 2^X$ with $\mathrm{graph}(C) \in \mathscr{B} \otimes \mathscr{F}$ is a closed random set, provided the σ-algebra $\mathscr{F}$ is universally complete. But this is immediate from the projection theorem 2.12, since for every open $U \subset X$

$$\{\omega : C(\omega) \cap U \neq \emptyset\} = \pi_\Omega\Big(\mathrm{graph}(C) \cap (U \times \Omega)\Big) \in \mathscr{F}$$

which is condition (ii) of Proposition 2.4. $\square$

Tightness of Compact Random Sets

We will need the standard fact that a compact metric space is characterised by being complete and totally bounded.

2.14 Theorem *A metric space (Y, d) is compact if, and only if, it is complete and totally bounded; i.e., for every $\delta > 0$ there is a finite set $Z \subset Y$ such that, for every $y \in Y$, there is some $z \in Z$ with $d(y, z) < \delta$.*

For a proof see, e.g., Dudley [19], Theorem 2.3.1, p. 35.

The proof of the following proposition is essentially the same as that of Ulam's Theorem (which states that on a Polish space any finite Borel measure is regular; see Dudley [19], Theorem 7.1.4, p. 176).

2.15 Proposition *Suppose that $\omega \mapsto K(\omega)$ is a compact random set. Then for every $\varepsilon > 0$ there exists a (non-random) compact set $K_\varepsilon \subset X$ such that $P\{\omega : K(\omega) \subset K_\varepsilon\} \geq 1 - \varepsilon$.*

PROOF Let $\{x_k : k \in \mathbb{N}\}$ be a countable dense subset of X. Then $X = \bigcup_{k=0}^{\infty} \overline{B}(x_k, \delta)$ for any $\delta > 0$, and hence, for any $m \in \mathbb{N}$,

$$\lim_{n \to \infty} P\left\{\omega : K(\omega) \subset \bigcup_{k=0}^{n} \overline{B}(x_k, 1/m)\right\} = 1,$$

where we used that

$$\left\{\omega : K(\omega) \subset \bigcup_{0}^{n} \overline{B}(x_k, 1/m)\right\} = \left\{\omega : K(\omega) \cap \left(\bigcup_{0}^{n} \overline{B}(x_k, 1/m)\right)^c = \emptyset\right\}$$

is measurable by Proposition 2.4 (ii). Given $m \in \mathbb{N}$ and $\varepsilon > 0$ choose $n = n(m, \varepsilon)$ big enough, such that

$$P\left\{\omega : K(\omega) \subset \bigcup_{k=0}^{n} \overline{B}(x_k, 1/m)\right\} \geq 1 - \frac{\varepsilon}{2^m}$$

and put $K_\varepsilon = \bigcap_{m \in \mathbb{N}} K(m, \varepsilon)$, where $K(m, \varepsilon) = \bigcup_{k=0}^{n(m,\varepsilon)} \overline{B}(x_k, 1/m)$. Then

$$\begin{aligned}
P\{K(\omega) \not\subset K_\varepsilon\} &= P\{K(\omega) \cap K_\varepsilon^c \neq \emptyset\} \\
&= P\left\{K(\omega) \cap \bigcup_{m \in \mathbb{N}} K(m, \varepsilon)^c \neq \emptyset\right\} \\
&= P\left\{\bigcup_{m \in \mathbb{N}} (K(\omega) \cap K(m, \varepsilon)^c) \neq \emptyset\right\} \\
&\leq \sum_{m \in \mathbb{N}} P\{K(\omega) \not\subset K(m, \varepsilon)\} \\
&\leq \sum_{m \in \mathbb{N}} \frac{\varepsilon}{2^m} \\
&= \varepsilon.
\end{aligned}$$

Since $K(m, \varepsilon)$ is closed for each m, also K_ε is closed, thus complete. Furthermore, for any $m \in \mathbb{N}$, $K_\varepsilon \subset K(m, \varepsilon) = \bigcup_{k=0}^{n(m,\varepsilon)} \overline{B}(x_k, 1/m)$, and hence K_ε is totally bounded. Consequently, K_ε is compact by Theorem 2.14. $\square$

2.16 Remark For compact random subsets of a Polish space X the notion of measurability can be formulated in terms of the *Hausdorff metric*, which is given by $d_H(K, M) = \max\{\sup\{d(k, M) : k \in K\},\ \sup\{d(m, K) : m \in M\}\}$ for $K, M \subset X$ compact. The space of compact subsets of a Polish space is Polish itself with respect to the Hausdorff metric, see Castaing and Valadier [9], Corollary II.9, p. 43. A map K from Ω to this space is measurable with respect to the Borel σ-algebra associated with the Hausdorff metric if and only if $\{\omega : K(\omega) \cap U \neq \emptyset\}$ is measurable for every open $U \subset X$, see Castaing and Valadier [9], Theorem II.10, p. 43. This means that K is a closed random set which is compact. In particular, any compact random set is a random variable taking values in this Polish space, so its distribution is tight.

Comments

The aim of the brief introduction to random sets given here is to provide the basic properties needed for the study of random measures. We like to stress that we did not assume the measurable structure to be given by a complete probability space. We therefore encounter problems which do not appear as soon as one assumes a complete probability space.

Basic monographs on set valued analysis in general and on measurable set mappings in particular are the books by Matheron [30] and by Aubin and Frankowska [2].

Random sets are of importance in other areas of mathematics too. In particular, random geometry uses random sets in spaces with geometrical structures. The large variety of phenomena and properties of random sets investigated in stochastic geometry needs considerably more structure of the state space than used here, where the state space is only assumed to be a Polish space, so not even a metric is given.

Chapter 3

Random Probability Measures and the Narrow Topology

Throughout this chapter $(\Omega, \mathscr{F}, P)$ is a probability space, and X is a Polish space equipped with a complete metric d. The σ-algebra of Borel sets of X is denoted by $\mathscr{B}$. The product space $X \times \Omega$ is understood to be a measurable space with the product σ-algebra $\mathscr{B} \otimes \mathscr{F}$, which is the smallest σ-algebra on $X \times \Omega$ with respect to which both the canonical projections $\pi_X : X \times \Omega \to X$ and $\pi_\Omega : X \times \Omega \to \Omega$ are measurable.

Random Probability Measures: Definition and Basic Properties

3.1 Definition A map

$$
\begin{aligned}
\mu : \mathscr{B} \times \Omega &\to [0,1], \\
(B, \omega) &\mapsto \mu_\omega(B),
\end{aligned}
$$

satisfying

  (i) for every $B \in \mathscr{B}$, $\omega \mapsto \mu_\omega(B)$ is measurable,

  (ii) for P-almost every $\omega \in \Omega$, $B \mapsto \mu_\omega(B)$ is a Borel probability measure,

is said to be a *random probability measure on* X, and is denoted by $\omega \mapsto \mu_\omega$.

Random probability measures are well known under several names such as *transition probabilities* or *Markov kernels*. Since we will be dealing with

17

probability measures in the following we just speak of measures, in particular
of random measures. See Remark 3.20 for some motivation for the choice of
the notion 'random measures' in the present context.

3.2 Remark If $\mu : \mathscr{B} \times \Omega \to [0,1]$ satisfies (ii) from Definition 3.1, and if $\omega \mapsto$
$\mu_\omega(D)$ is measurable for every D from a $\cap$-stable family $\mathcal{E}$ of Borel subsets
of X which generate $\mathscr{B}$ (i.e., $\sigma(\mathcal{E}) = \mathscr{B}$), then (i) is satisfied as well, so μ is
a random measure. In fact, $\mathcal{D} = \{D \in \mathscr{B} : \omega \mapsto \mu_\omega(D) \text{ is measurable}\}$ is a
Dynkin system with $\mathcal{E} \subset \mathcal{D}$. Since $\cap$-stability of $\mathcal{E}$ implies that the σ-algebra
generated by $\mathcal{E}$ coincides with the smallest Dynkin system containing $\mathcal{E}$,
$\mathscr{B} = \sigma(\mathcal{E}) \subset \mathcal{D}$, hence $\mathscr{B} = \mathcal{D}$.
Consequently, it suffices to have (ii) together with measurability of $\omega \mapsto$
$\mu_\omega(U)$ for all open or for all closed $U \subset X$, respectively, to conclude that μ
is a random measure.

We collect some standard results.

3.3 Proposition *Suppose that $\omega \mapsto \mu_\omega$ is a random measure. Then*

(i) *for all measurable $h : X \times \Omega \to \mathbb{R}$ with h bounded or nonnegative the
map*

$$\omega \mapsto \int_X h(x,\omega)d\mu_\omega(x)$$

is measurable.

(ii) *The assignment*

$$A \mapsto \int_\Omega \int_X 1_A(x,\omega)d\mu_\omega(x)dP(\omega)$$

*$A \in \mathscr{B} \otimes \mathscr{F}$, defines a probability measure on $X \times \Omega$, which is denoted
by μ. The marginal of μ on Ω is P, i.e., $\pi_\Omega\mu = P$.*

PROOF (i) Put

$$\mathcal{H} = \Big\{ h : X \times \Omega \to \mathbb{R} : h \text{ bounded measurable,}$$
$$\omega \mapsto \int_X h(x,\omega)d\mu_\omega(x) \text{ measurable} \Big\}.$$

Then $\mathcal{H}$ is a vector space with $1_A \in \mathcal{H}$ for all $A \in \mathscr{B} \times \mathscr{F}$. Furthermore, if $0 \le$
$h_n \in \mathcal{H}$, $n \in \mathbb{N}$, with $h_n \nearrow h$ for some bounded h, then $h \in \mathcal{H}$ by monotone
convergence. By a monotone class argument (Williams [39], Theorem II.4,
p. 40), $\mathcal{H}$ contains all bounded $\sigma(\mathscr{B} \times \mathscr{F})$-measurable functions.
For h nonnegative put $h_n = h \wedge n = \min\{h, n\}$. Then h_n is bounded and
measurable, and thus $\omega \mapsto \int h_n(\cdot, \omega)\, d\mu_\omega$ is measurable. Since $h_n \nearrow h$,

the monotone convergence theorem yields convergence of $\int h_n(\cdot,\omega)\,d\mu_\omega$ to $\int h(\cdot,\omega)\,d\mu_\omega$.

(ii) See Gänssler and Stute [24], Satz 1.8.10, p. 44.   □

3.4 Corollary *Suppose that* $\omega \mapsto A(\omega)$ *is a set valued map such that* graph(A) *is a measurable subset of* $X \times \Omega$. *Then* $\omega \mapsto \mu_\omega(A(\omega))$ *is well defined and measurable.*

PROOF Measurability of graph(A) implies that all ω-sections $A(\omega)$ are Borel sets, so $\mu_\omega(A(\omega))$ is well defined. Measurability follows from Proposition 3.3 (i) with $h(x,\omega) = 1_{A(\omega)}(x)$, which is the indicator function of graph(A).   □

Measurability of graph(A) is precisely condition (iv) of Proposition 2.4. Consequently, measurability of $\omega \mapsto \mu_\omega(A(\omega))$ for any closed random set A holds as well as measurability of $\omega \mapsto \mu_\omega(A(\omega)) = 1 - \mu_\omega(A^c(\omega))$ for any open random set A.

3.5 Proposition *If* μ *and* ν *are probability measures on* $(X \times \Omega, \mathscr{B} \otimes \mathscr{F})$ *such that* $\mu(B \times F) = \nu(B \times F)$ *for all closed (or for all open, respectively)* $B \subset X$ *and for all* $F \in \mathscr{F}_0$, *where* $\mathscr{F}_0 \subset \mathscr{F}$ *is closed under intersections and generates* $\mathscr{F}$ *(i.e.,* $\sigma(\mathscr{F}_0) = \mathscr{F}$*), then* $\mu = \nu$ *on* $\mathscr{B} \otimes \mathscr{F}$.

PROOF The set system $\{B \times F : B \subset X \text{ closed}, F \in \mathscr{F}_0\}$ is closed under intersections and generates $\mathscr{B} \otimes \mathscr{F}$. Hence equality of μ and ν on $\mathscr{B} \otimes \mathscr{F}$ follows by a standard monotone class argument (Gänssler and Stute [24], Satz 1.4.10, p. 28).   □

Next consider the space of all probability measures on $(X \times \Omega, \mathscr{B} \otimes \mathscr{F})$ with marginal P on Ω. The connection between marginal P measures on the product space $X \times \Omega$ and random measures is given by the following proposition.

3.6 Proposition (EXISTENCE AND UNIQUENESS OF A DISINTEGRATION) *For every probability measure* μ *on* $X \times \Omega$ *with* $\pi_\Omega\mu = P$ *there exists a random measure* $\omega \mapsto \mu_\omega$ *such that*

$$\int_{X\times\Omega} h(x,\omega)d\mu(x,\omega) = \int_\Omega \int_X h(x,\omega)d\mu_\omega(x)dP(\omega) \qquad (3.1)$$

for every bounded measurable $h : X \times \Omega \to \mathbb{R}$. *The random measure* $\omega \mapsto \mu_\omega$ *is unique* P-a.s.

PROOF   Existence of $\omega \mapsto \mu_\omega$ follows from Satz 5.3.21 of Gänssler and
Stute [24], p. 198. To prove uniqueness, suppose that $\omega \mapsto \mu_\omega$ and $\omega \mapsto \nu_\omega$
both satisfy (3.1). Then $\mu_\omega(B) = \nu_\omega(B)$ for P-almost all ω for any $B \in \mathscr{B}$.
Since X is Polish, its topology is countably generated, and outside a P-
nullset μ_ω and ν_ω coincide on all sets of a $\cap$-stable countable generator of the
topology. Thus outside this nullset μ_ω and ν_ω coincide on the σ-algebra generated by the $\cap$-stable generator of the topology by Satz 1.4.10 of Gänssler
and Stute [24], p. 28, which is $\mathscr{B}$ (see [24], Satz 1.1.28, p. 15).   □

The random measure $\omega \mapsto \mu_\omega$ associated with any $\mu \in Pr(X \times \Omega)$ with
marginal P on Ω given by Proposition 3.6 is often addressed to as the *disin-*
tegration of μ. We will identify two random measures $\omega \mapsto \mu$ and $\omega \mapsto \nu$ if
$\mu_\omega = \nu_\omega$ for P-almost all ω. Put

$$Pr_\Omega(X) = \big\{\mu : \mathscr{B} \times \Omega \to [0,1] : \omega \mapsto \mu_\omega \text{ random measure}\big\}$$

with two random measures identified if they coincide P-a.s., and

$$Pr_P(X \times \Omega) = \{\mu \in Pr(X \times \Omega) : \pi_\Omega \mu = P\}.$$

Both $Pr_\Omega(X)$ and $Pr_P(X \times \Omega)$ carry a canonical convex structure in the sense
that for any two random measures μ and ν and any $p \in [0,1]$ also $p\mu + (1-p)\nu$
is a random measure, given by $(p\mu_\omega + (1-p)\nu_\omega)(B) = p\mu_\omega(B) + (1-p)\nu_\omega(B)$.
For $Pr_P(X \times \Omega)$ the convex structure is simply given by pointwise addition.
By Propositions 3.3 (ii) and 3.6 we get an isomorphism between $Pr_\Omega(X)$ and
$Pr_P(X \times \Omega)$, given by (3.1), or equivalently by

$$\mu(A) = \int_\Omega \int_X 1_A(x,\omega) d\mu_\omega(x) dP(\omega)$$

for all $A \in \mathscr{B} \otimes \mathscr{F}$. The convex structure is preserved by this isomorphism.
We will usually identify $Pr_P(X \times \Omega)$ and $Pr_\Omega(X)$ via these relations without
further notice. Henceforth we will only speak of a random measure $\mu \in$
$Pr_\Omega(X)$. It should become clear from the respective context when μ is to
be understood as a transition probability from Ω to X, and when it is to be
understood as a measure on the product space $X \times \Omega$.

In particular, for $\omega \to A(\omega)$ a random set (in the sense that $\text{graph}(A) =$
$\{(x,\omega) : x \in A(\omega)\}$ is in $\mathscr{B} \otimes \mathscr{F}$), we sometimes identify A with $\text{graph}(A)$
notationally and write

$$\mu(A) = \mu(\text{graph}(A)) = \int_\Omega \mu_\omega(A(\omega)) \, dP(\omega).$$

3.7 Definition Suppose that μ is a random measure.

(i) The *support of* μ is (the set valued mapping) $\omega \mapsto \operatorname{supp} \mu_\omega$. (Recall that the support of a Borel measure is the complement of the union of all open sets with μ_ω-measure zero, hence it is closed.)

(ii) If $\omega \mapsto B(\omega)$ is a random set, then μ is said to be *supported by B* if $\mu_\omega(B(\omega)) = 1$ P-a.s. (or, equivalently, if $\mu(\operatorname{graph}(B)) = 1$).

3.8 Remark (i) The support of a random measure is a closed random set (i.e., it is measurable). In fact, let ρ be any Borel probability measure on X, and let $U \subset X$ be open. Then $\rho(U) > 0$ if and only if $U \cap \operatorname{supp} \rho \neq \emptyset$. Consequently, if μ is a random measure, and $U \subset X$ is open, then

$$\{\omega : \operatorname{supp} \mu_\omega \cap U \neq \emptyset\} = \{\omega : \mu_\omega(U) > 0\}$$

which is measurable by (i) of Definition 3.1. Measurability of $\omega \mapsto \operatorname{supp} \mu_\omega$ follows from Proposition 2.4.

(ii) For $A_1, A_2 \in \mathscr{B} \otimes \mathscr{F}$ we have $\mu(A_1) = \mu(A_2)$ for every random measure μ if and only if $A_1(\omega) = A_2(\omega)$ for P-almost all $\omega \in \Omega$.

For $\mu \in Pr_\Omega(X)$ and a bounded measurable (or μ-integrable) $h : X \times \Omega \to \mathbb{R}$ we write

$$\mu(h) = \int_\Omega \int_X h(x, \omega) d\mu_\omega(x) dP(\omega) = \int_{X \times \Omega} h(x, \omega) d\mu(x, \omega). \qquad (3.2)$$

Random Continuous Functions

Now let $f : X \times \Omega \to \mathbb{R}$ be a function such that

(i) for all $x \in X$ the x-section $\omega \mapsto f(x, \omega)$ is measurable,

(ii) for all $\omega \in \Omega$ the ω-section $x \mapsto f(x, \omega)$ is continuous and bounded,

(iii) $\omega \mapsto \sup\{|f(x, \omega)| : x \in X\}$ is integrable with respect to P (it is measurable by separability of X).

Any such f is jointly measurable by Lemma 1.1. If f and g are functions both satisfying (i)–(iii) then $\{\omega : f(\cdot, \omega) \neq g(\cdot, \omega)\}$ is measurable again by separability of X. Identify f and g if $P\{f(\cdot, \omega) \neq g(\cdot, \omega)\} = 0$. By the common misuse of notation the equivalence class of f will be denoted by f again.

3.9 Definition　A *random continuous function* is (the equivalence class of) a function $f : X \times \Omega \to \mathbb{R}$ satisfying (i)–(iii) above. The set of all random continuous functions is a linear space, denoted by $C_\Omega(X)$.

A norm on $C_\Omega(X)$ is defined by

$$|f|_\infty = \int \sup_{x \in X} |f(x, \omega)| \, dP(\omega).$$

Note that the norm $| \cdot |_\infty$ should rather be denoted $| \cdot |_{\infty \times 1}$ to notify that it means supremum on X and $L^1(P)$ on Ω. In particular, $| \cdot |_\infty$ should not be confused with the $L^\infty(P)$-norm on the real valued functions on Ω.

It would be more precise to speak of (*P-a.s.*) *bounded* random continuous functions, and to denote the space by $C_\Omega^b(X)$ instead of $C_\Omega(X)$. For the sake of brevity of notation we rather use $C_\Omega(X)$.

Random continuous functions have been used by Balder [3], Appendix A, p. 591, and by Valadier [37], p. 154, under the name 'Carathéodory integrands'.

There are two particular classes of random continuous functions. Firstly, if $g : X \to \mathbb{R}$ is a bounded continuous function, then (the equivalence class of) $(x, \omega) \mapsto g(x)$ defines an element of $C_\Omega(X)$. We will sometimes call $f \in C_\Omega(X)$ a *non-random* element if there is a bounded continuous $g : X \to \mathbb{R}$ with $f(\cdot, \omega) = g(\cdot)$ for *P*-almost all ω. (Note that for $f \in C_\Omega(X)$ by separability of X it suffices to have $f(x, \cdot) = g(x)$ *P*-a.s. for every x from a countable dense set in X to get $f(\cdot, \omega) = g(\cdot)$ *P*-a.s.) Secondly, let $h : \Omega \to \mathbb{R}$ be (the equivalence class of) a function which is integrable with respect to P, commonly denoted as $h \in L^1(P)$. Then h can be considered as the random continuous function given by $(x, \omega) \mapsto h(\omega)$. In the following, elements of $L^1(P)$ will be understood as random continuous functions without further mentioning. Clearly $\mu(h) = \int h \, dP$, independent of $\mu \in Pr_\Omega(X)$, for any $h \in L^1(P)$.

For any $g : X \to \mathbb{R}$ bounded continuous and any $h \in L^1(X)$, the assignment $f(x, \omega) = g(x) \, h(\omega)$ defines a random continuous function.

3.10 Remark　(i)　Suppose that $f : X \times \Omega \to \mathbb{R}$ is measurable in ω for every $x \in X$, and continuous in x for every $\omega \in \Omega$. Then f is measurable with respect to $\mathscr{B} \otimes \mathscr{F}$ by Lemma 1.1. For $r \in \mathbb{R}$ thus the level sets

$$F_r(\omega) = \{x : f(x, \omega) \geq r\} \quad \text{and} \quad F^r(\omega) = \{x : f(x, \omega) \leq r\}$$

define closed set valued maps with measurable graph, hence they are closed random sets with respect to the universally completed σ-algebra of $\mathscr{F}$ by Proposition 2.4.

(ii) Suppose again that $f : X \times \Omega \to \mathbb{R}$ is measurable in ω for every $x \in X$, and continuous in x for every $\omega \in \Omega$. Let $\omega \mapsto C(\omega)$ be a nonvoid closed random set. Then

$$\omega \mapsto \sup_{x \in C(\omega)} f(x, \omega) \tag{3.3}$$

is measurable (without completion, which was needed for Corollary 2.13). In fact, by Theorem 2.6 there exists a sequence $c_n : \Omega \to X$, $n \in \mathbb{N}$, of measurable selections of C with $C(\omega) = \mathrm{cl}\{c_n(\omega) : n \in \mathbb{N}\}$. Continuity of $x \mapsto f(x, \omega)$ implies

$$\sup_{x \in C(\omega)} f(x, \omega) = \sup_{n \in \mathbb{N}} f(c_n(\omega), \omega).$$

If C is void with positive probability then still (3.3) yields a measurable map from Ω to $\overline{\mathbb{R}}$ (with $\sup \emptyset = -\infty$).

(iii) Suppose that f is a random continuous function. Then dominated convergence implies that $F(x) = \int f(x, \omega)\, dP(\omega)$ is a continuous function on X, which is bounded by $\int \sup_x f(x, \omega)\, dP(\omega)$ (see, e.g., Bauer [6], Lemma 16.1, p. 101).

Let μ be a random measure, and let f be a random continuous function. Since each random continuous function is jointly measurable by Lemma 1.1, the integral

$$\mu(f) = \int_\Omega \int_X f(x, \omega) d\mu_\omega(x) dP(\omega) = \int_{X \times \Omega} f(x, \omega)\, d\mu(x, \omega)$$

is well defined (it depends neither on the choice of μ nor on the choice of f from the respective equivalence classes). Furthermore, $f \mapsto \mu(f)$ is linear for each random measure μ, and $\mu \mapsto \mu(f)$ is convex for each random continuous function f.

The proof of the following lemma is immediate.

3.11 Lemma *For each random measure μ and $f, g \in C_\Omega(X)$*

$$|\mu(f) - \mu(g)| \le |f - g|_\infty = \int_\Omega \sup_{x \in X} |f(x, \omega) - g(x, \omega)|\, dP(\omega).$$

In particular, the family of maps from $C_\Omega(X)$ to $\mathbb{R}$ given by $\{f \mapsto \mu(f) : \mu \in Pr_\Omega(X)\}$ is uniformly equicontinuous (with respect to the metric given by $|\cdot|_\infty$ on $C_\Omega(X)$).

We will use Lipschitz functions to topologise random probability measures on metric spaces. With this approach we follow Dudley [19], Chapter 11.

3.12 Definition   A function $g : X \to \mathbb{R}$ is *Lipschitz* if

$$[g]_\mathrm{L} = \sup_{x \neq y} \frac{|g(x) - g(y)|}{d(x, y)} < \infty. \tag{3.4}$$

With $\|g\|_\mathrm{BL} = \max\{[g]_\mathrm{L}, \sup_{x \in X} |g(x)|\}$ the space

$$\mathrm{BL}(X) = \mathrm{BL}(X, d) = \{g : X \to \mathbb{R} : \|g\|_\mathrm{BL} < \infty\}$$

is a Banach space. Furthermore, $\mathrm{BL}(X)$ is a subspace of the bounded continuous functions from X to $\mathbb{R}$. Elements of $\mathrm{BL}(X)$ are called *bounded Lipschitz* functions on X.

3.13 Remark   (i)   The space of bounded Lipschitz functions is even an algebra. In fact, for $g_1, g_2 \in \mathrm{BL}(X)$ and $x, y \in X$

$$
\begin{aligned}
|g_1(x)g_2(x) - g_1(y)g_2(y)| &\leq |g_1(x)(g_2(x) - g_2(y))| + |g_2(y)(g_1(x) - g_1(y))| \\
&\leq \left(|g_1|_\infty [g_2]_\mathrm{L} + |g_2|_\infty [g_1]_\mathrm{L}\right) d(x, y),
\end{aligned}
$$

hence $g_1 g_2 \in \mathrm{BL}(X)$ and $\|g_1 g_2\|_\mathrm{BL} \leq 2\|g_1\|_\mathrm{BL} \|g_2\|_\mathrm{BL}$. The above argument is due to Dudley [19], Proposition 11.2.1, p. 306. The present definition of $\|\cdot\|_\mathrm{BL}$ differs from that of Dudley (who defines it as $\sup|g| + [g]_\mathrm{L}$). This will be slightly more convenient later. The present definition has the disadvantage that $\|\cdot\|_\mathrm{BL}$ is not compatible with the algebra structure.

(ii)   For any $g_1, \ldots, g_n : X \to \mathbb{R}$, both $[\min\{g_k : 1 \leq k \leq n\}]_\mathrm{L}$ as well as $[\max\{g_k : 1 \leq k \leq n\}]_\mathrm{L}$ are bounded by $\max\{[g_k]_\mathrm{L} : 1 \leq k \leq n\}$. Both $\|\min\{g_k : 1 \leq k \leq n\}\|_\mathrm{BL}$ and $\|\max\{g_k : 1 \leq k \leq n\}\|_\mathrm{BL}$ are bounded by $2\max\{\|g_k\|_\mathrm{BL} : 1 \leq k \leq n\}$, see Dudley [19], Proposition 11.2.2, p. 307.

(iii)   For any $x, y \in X$ and any $B \subset X$

$$d(x, B) \leq d(x, y) + d(y, B),$$

hence $x \mapsto d(x, B)$ is a Lipschitz function with $[d(\cdot, B)]_\mathrm{L} \leq 1$ for arbitrary $B \subset X$.

3.14 Lemma   *If for μ and $\nu \in Pr_\Omega(X)$*

$$\int_F \mu_\omega(g) \, dP(\omega) = \int_F \nu_\omega(g) \, dP(\omega)$$

for all $g \in \mathrm{BL}(X)$ with $0 \leq g \leq 1$ and for $F \in \mathscr{F}_0$, for some $\mathscr{F}_0 \subset \mathscr{F}$ which is closed under intersections and satisfies $\sigma(\mathscr{F}_0) = \mathscr{F}$, then $\mu = \nu$. Here

$\mu_\omega(g) = \int_X g(x)\,d\mu_\omega(x)$, ditto for $\nu_\omega(g)$. In particular, if $\mu(f) = \nu(f)$ for all random continuous functions f with $0 \le f \le 1$, then $\mu = \nu$.

PROOF By Proposition 3.5 it suffices to prove $\mu(B \times F) = \nu(B \times F)$ for all closed $B \in \mathcal{B}$ and $F \in \mathcal{F}_0$. For $n \in \mathbb{N}$ put $b_n(x) = 1 - \min\{n\,d(x, B), 1\}$. Then $b_n \in \mathrm{BL}(X)$, and $0 \le b_n \le 1$. Furthermore, $f_n(x, \omega) = b_n(x)1_F(\omega)$ converges to $1_B 1_F$ monotonically. By monotone convergence $\mu(B \times F) = \lim \mu(f_n) = \lim \nu(f_n) = \nu(B \times F)$, where the limit is taken for $n \to \infty$. $\qquad\square$

Definition of The Narrow Topology

3.15 Definition The topology on $Pr_\Omega(X)$ generated by the functions $\mu \mapsto \mu(f)$, $f \in C_\Omega(X)$, is called the *narrow topology on $Pr_\Omega(X)$*.

A neighbourhood basis for the narrow topology in $\nu \in Pr_\Omega(X)$ is given by

$$U_{f_1,\dots,f_n;\delta}(\nu) = \{\mu \in Pr_\Omega(X) : |\textstyle\int f_k\,d\mu - \int f_k\,d\nu| < \delta,\ k = 1,\dots,n\}, \quad (3.5)$$

where $n \in \mathbb{N}$, $f_1, \dots, f_n \in C_\Omega(X)$, and $\delta > 0$.

As an immediate consequence of Lemma 3.14 we note that the narrow topology is Hausdorff. We continue by proving that the narrow topology is generated by $f \in C_\Omega(X)$ with $0 \le f \le 1$.

3.16 Lemma *If for some topology on $Pr_\Omega(X)$ (not necessarily the narrow topology) $\mu \mapsto \mu(f)$ is continuous for every $f \in C_\Omega(X)$ with $P\{\omega : 0 \le f(\cdot, \omega) \le 1\} = 1$, then $\mu \mapsto \mu(f)$ is continuous in this topology for every $f \in C_\Omega(X)$.*

PROOF Pick $f \in C_\Omega(X)$ with $f \ge 0$ P-a.s., and let $\varepsilon > 0$. Choose $N \in \mathbb{N}$ such that

$$\int_{F_N} \sup_{x \in X} f(x, \omega)\,dP(\omega) < \varepsilon/3,$$

where $F_N = \{\omega : \sup_{x \in X} f(x, \omega) \ge N\}$ (existence of such an N follows from integrability of $\omega \mapsto \sup_{x \in X} f(x, \omega)$). Then

$$|f - f \wedge N|_\infty \le \int_{F_N} \sup_{x \in X} |f(x, \omega)|\,dP(\omega) < \varepsilon/3, \qquad (3.6)$$

where $|\cdot|_\infty$ is the norm on $C_\Omega(X)$ introduced in Definition 3.9, and $f \wedge N = \min\{f, N\}$. Fix $\mu \in Pr_\Omega(X)$, and put

$$
\begin{aligned}
U &= \{\nu \in Pr_\Omega(X) : |\mu(f \wedge N) - \nu(f \wedge N)| < \varepsilon/3\} \\
&= \{\nu : |\mu((f \wedge N)/N) - \nu((f \wedge N)/N)| < \varepsilon/3N\},
\end{aligned}
$$

where $f \wedge N = \min\{f, N\}$. Since $0 \le (f \wedge N)/N \le 1$, U is an open neighbourhood of μ in the topology under consideration. From Lemma 3.11 together with (3.6) we get for any $\nu \in U$

$$\begin{aligned}|\mu(f) - \nu(f)| &\le |\mu(f) - \mu(f \wedge N)| + |\mu(f \wedge N) - \nu(f \wedge N)| \\ &\quad + |\nu(f \wedge N) - \nu(f)| \\ &< 3\frac{\varepsilon}{3} = \varepsilon.\end{aligned}$$

This holds for $\mu \in Pr_\Omega(X)$ arbitrary. Consequently, $\mu \mapsto \int f \, d\mu$ is continuous for every $f \in C_\Omega(X)$ with $f \ge 0$. Finally, for $f \in C_\Omega(X)$ arbitrary the assertion follows from continuity of $\mu \mapsto \mu(f^+) - \mu(f^-) = \mu(f)$.   □

Lemma 3.16 implies that the narrow topology coincides with the topology generated by $\mu \mapsto \mu(f)$, $f \in C_\Omega(X)$, with $0 \le f \le 1$ P-a.s.

The Portmenteau Theorem

Next we are concerned with the Portmenteau Theorem for random measures. The narrow topology will turn out not to be metrisable even for the case where X is the unit interval in $\mathbb{R}$, say, unless the σ-algebra $\mathscr{F}$ is countably generated (mod P). So it is meaningful to state the theorem for generalised sequences, i.e., for nets. The proof remains the same.

3.17 Theorem (PORTMENTEAU THEOREM) *Let $\{\mu^\alpha\}$ be a net in $Pr_\Omega(X)$. Then the following statements are equivalent:*

(i) *μ^α converges to μ in the narrow topology.*

(ii) *$\limsup_\alpha \mu^\alpha(C) \le \mu(C)$ for all closed random sets C.*

(iii) *$\limsup_\alpha \mu^\alpha(C^0) \le \mu(C^0)$ for all closed random sets C^0 with respect to the P-completion $\mathscr{F}^0$ of $\mathscr{F}$.*

(iv) *$\liminf_\alpha \mu^\alpha(U) \ge \mu(U)$ for all open random sets U.*

(v) *$\liminf_\alpha \mu^\alpha(U) \ge \mu(U)$ for all open random sets U^0 with respect to $\mathscr{F}^0$.*

PROOF   Clearly (iii) implies (ii), and (v) implies (iv). Equivalence of (ii) and (iv) as well as equivalence of (iii) and (v) is obvious by taking complements. To see that (ii) implies (iii), let C^0 be a closed random set with respect to $\mathscr{F}^0$. By Lemma 2.7 there exists a closed random set C with respect to the uncompleted $\mathscr{F}$ with $C = C^0$ P-a.s. Since this implies $\nu(C^0) = \nu(C)$ for all random measures ν by Remark 3.8 (ii), it follows that (ii) implies (iii).

To prove that (i) implies (ii), let C be a closed random set. For $k \in \mathbb{N}$ put $f_k(x,\omega) = 1 - \min\{k\,d(x,C(\omega)),1\}$. Then $f_k \in C_\Omega(X)$, $f_k \geq 1_C$, and f_k decreases monotonically to 1_C with $k \to \infty$. For each fixed k

$$\limsup_\alpha \mu^\alpha(C) \leq \lim_\alpha \mu^\alpha(f_k) = \mu(f_k),$$

hence

$$\limsup_\alpha \mu^\alpha(C) \leq \inf_k \mu(f_k) = \mu(C).$$

We finish the proof by showing that (iii) implies (i). By Lemma 3.16 it suffices to prove $\lim_\alpha \mu^\alpha(f) = \mu(f)$ for every random continuous function f with $0 \leq f \leq 1$. We first establish

$$\limsup_\alpha \mu^\alpha(f) \leq \mu(f). \tag{3.7}$$

Fix $n \in \mathbb{N}$ and define $C_k \subset X \times \Omega$, $0 \leq k \leq n$, by

$$C_k(\omega) = \left\{x \in X : f(x,\omega) \geq \frac{k}{n}\right\}.$$

Then C_k is a closed random set with respect to the universally completed σ-algebra associated with $\mathscr{F}$ by Remark 3.10 (i), $0 \leq k \leq n$. The universally complete σ-algebra associated with $\mathscr{F}$ is contained in the completion of $\mathscr{F}$ with respect to P, so (iii) applies to C_k, $0 \leq k \leq n$. Using Lemma 1.4 twice (ω-wise, once for μ and once for μ^α), we get from (iii)

$$\mu(f) \geq \frac{1}{n}\sum_{k=1}^{n}\mu(C_k) \geq \frac{1}{n}\sum_{k=1}^{n}\limsup_\alpha \mu^\alpha(C_k) \geq \limsup_\alpha \frac{1}{n}\sum_{k=1}^{n}\mu^\alpha(C_k)$$

$$= \limsup_\alpha \frac{1}{n}\Big(\sum_{k=0}^{n}\mu^\alpha(C_k) - 1\Big) \geq \limsup_\alpha \mu^\alpha(f) - \frac{1}{n}.$$

Since n is arbitrary we obtain (3.7) for any $f \in C_\Omega(X)$ with $0 \leq f \leq 1$. Finally, applying (3.7) to $1 - f$ and using $\nu(1) = 1$ for every $\nu \in Pr_\Omega(X)$ we get

$$\mu(f) \leq \liminf_\alpha \mu^\alpha(f)$$

for any $f \in C_\Omega(X)$ with $0 \leq f \leq 1$, hence the assertion follows.   □

3.18 Corollary *Suppose that C is a closed random set. Then the set of all random measures supported by C,*

$$\Gamma = \{\mu \in Pr_\Omega(X) : \mu_\omega(C(\omega)) = 1 \ P\text{-a.s.}\},$$

is closed in the narrow topology. This applies, in particular, for any closed non-random $C \subset X$.

PROOF Suppose that $(\mu^\alpha)_\alpha$ is a net in Γ converging to $\mu \in Pr_\Omega(X)$ in the narrow topology. Thus (i) of Theorem 3.17 is satisfied, and so (ii), applied with the closed random set C, yields $1 = \limsup_\alpha \mu^\alpha(C) \le \mu(C)$. Consequently, also μ is supported by C.   □

The next lemma will be needed for random dynamical systems later.

3.19 Lemma *Suppose that C is a closed random set in a Polish space X, over $(\Omega, \mathscr{F}, P)$. Let $f : X \times \Omega \to \mathbb{R}$ be measurable, and assume that, for almost all ω, $x \mapsto f(x,\omega)$ is continuous on $C(\omega)$. Suppose further that*

$$\omega \mapsto \sup_{x \in C(\omega)} f^+(x,\omega) \tag{3.8}$$

is integrable with respect to P, and that, for P-almost all ω, the restriction of $x \mapsto f(x,\omega)$ to $C(\omega)$ is bounded from below by some $k(\omega)$, say. Then the map $\gamma \mapsto \int f \, d\gamma$ is well defined on $Pr_\Omega(X)$, possibly taking the value $-\infty$. Furthermore, this map is upper semi-continuous on the set of random probability measures supported by C.

PROOF For any random probability measure γ with $\gamma(C) = 1$, $\int f \, d\gamma$ is meaningful by (3.8), being either a real or equal to $-\infty$. Considering, instead of f, the function

$$(x,\omega) \mapsto \begin{cases} f(x,\omega) & \text{if } x \in C(\omega), \\ \sup_{z \in C(\omega)} \left(\dfrac{f(z,\omega) - \inf_{y \in C(\omega)} f(y,\omega)}{(1 + d(z,x)^2)^{1/d(x,C(\omega))}} \right) + \inf_{y \in C(\omega)} f(y,\omega) & \text{else,} \end{cases}$$

if necessary (compare Dudley [19], Exercise 7, p. 49), we may assume that f is P-a.s. continuous on X, and bounded by $\sup_{x \in C(\omega)} f(x,\omega)$ from above, without changing $\int f \, d\gamma$ for γ supported by C. Put $f_N = \max\{f, -N\}$ for $N \in \mathbb{N}$. Then, for every $N \in \mathbb{N}$, f_N is a random continuous function, whence $\gamma \mapsto \int f_N \, d\gamma$ is continuous on the space of random probability measures on X. Monotone convergence implies

$$\int f \, d\gamma = \lim_{N \to \infty} \int \max\{f, -N\} \, d\gamma = \inf_{N \in \mathbb{N}} \int \max\{f, -N\} \, d\gamma$$

for every random probability measure γ. The infimum of a family of continuous functions being upper semi-continuous, we get that on the (closed) set $\{\gamma \in Pr_\Omega(X) : \gamma(C) = 1\}$ the map $\gamma \mapsto \int f \, d\gamma$ is upper semi-continuous. □

The Narrow Topologies on Deterministic and on Random Measures

The narrow topology on $Pr_\Omega(X)$, the set of *random* probability measures, generalises the notion of narrow topology on $Pr(X)$, the set of *non-random* probability measures on X. Recall that the narrow topology on $Pr(X)$ is the coarsest topology on $Pr(X)$ such that $\rho \mapsto \int_X g \, d\rho$ is continuous for every $g : X \to \mathbb{R}$ bounded and continuous (Dellacherie and Meyer [18], III.54, p. 71; Williams [39], I.37, p. 22). Since X is Polish, $Pr(X)$ with the narrow topology is also Polish (Dellacherie and Meyer [18], III.60, p. 73). Denote by d a complete metric on $Pr(X)$ metrising the narrow topology. Note that X may be embedded into $Pr(X)$ by $x \mapsto \delta_x$, where δ_x denotes the Dirac measure in x. Thus $d(x, y) = d(\delta_x, \delta_y)$ gives a compatible metric on X, so it is justified to use the same letter for metrics on X and $Pr(X)$. Denote further by $C(X)$ the space of bounded continuous functions on X. As before, it should be denoted by $C_b(X)$ rather. For the sake of consistency of notation we stick to $C(X)$ for *bounded* continuous functions.

3.20 Remark (i) If μ is a random measure then $\omega \mapsto \mu_\omega$ is measurable with respect to the Borel σ-algebra of the narrow topology on $Pr(X)$. In fact, for any $g \in C(X)$ the map $\omega \mapsto \mu_\omega(g)$ is measurable by Proposition 3.3 (i). On the other hand, if $\mu : \Omega \to Pr(X)$ is measurable with respect to the Borel σ-algebra generated by the narrow topology on $Pr(X)$, then μ satisfies the conditions of Definition 3.1, so it is a random measure. In order to verify condition (i) of the definition, pick $B \subset X$ closed and put $b_n(x) = 1 - \min\{n \, d(x, B), 1\}$ for $n \in \mathbb{N}$. Then $b_n \in C(X)$, and b_n decreases to 1_B with $n \to \infty$, thus $\int b_n d\mu_\omega$ converges to $\mu_\omega(B)$ by monotone convergence for every $\omega \in \Omega$. Measurability of $\omega \to \mu_\omega(b_n)$, $n \in \mathbb{N}$, implies measurability of $\omega \mapsto \mu_\omega(B)$. This holds for every $B \subset X$ closed, hence, by Remark 3.2, for all $B \in \mathscr{B}$.

Consequently, random measures are precisely the measurable maps from Ω to $Pr(X)$, where $Pr(X)$ is equipped with the Borel σ-algebra of the narrow topology. This is the reason for addressing the objects defined in Definition 3.1 as random measures rather than transition probabilities or Markov kernels.

(ii) If μ and ν are random measures, continuity of $d : Pr(X) \times Pr(X) \to \mathbb{R}$ together with (i) imply measurability of $\omega \mapsto d(\mu_\omega, \nu_\omega)$. Here we need separability of the narrow topology of $Pr(X)$ to ensure that $\mathscr{B}(Pr(X)) \otimes \mathscr{B}(Pr(X)) = \mathscr{B}(Pr(X) \times Pr(X))$, where $\mathscr{B}(Pr(X) \times Pr(X))$ denotes the Borel σ-algebra of $Pr(X) \times Pr(X)$. See Gänssler and Stute [24], Satz 8.1.4 and Beispiel 8.1.5, pp. 329–330.

We now discuss the relations between the topological spaces $Pr(X)$ and $Pr_\Omega(X)$ equipped with their respective narrow topologies.

3.21 Remark Every non-random $\rho \in Pr(X)$ can be considered as an element of $Pr_\Omega(X)$ by $\omega \mapsto \rho$. This induces a map $e : Pr(X) \to Pr_\Omega(X)$, $\rho \mapsto (\omega \mapsto \rho)$. It is immediate that e is injective, so we may identify the space $Pr(X)$ of deterministic probability measures as the subset of $Pr_\Omega(X)$ consisting of those random measures $\omega \mapsto \mu_\omega$ which coincide with a non-random $\rho \in Pr(X)$ almost surely, i.e., $\mu_\omega = \rho$ for P-almost all ω. The narrow topology of $Pr(X)$ then coincides with the trace of the narrow topology of $Pr_\Omega(X)$ on $Pr(X)$, considering $Pr(X)$ as a subspace of $Pr_\Omega(X)$. In fact, the narrow topology on $Pr(X)$ is the topology making all $\rho \mapsto \rho(f)$ continuous, where f ranges through $C(X) \subset C_\Omega(X)$. So it is weaker than the trace of the narrow topology of $Pr_\Omega(X)$. On the other hand, for $f \in C_\Omega(X)$ and $\rho \in Pr(X)$ the theorem of Fubini yields

$$\int_\Omega \int_X f(x, \omega) \, d\rho(x) \, dP(\omega) = \int_X \int_\Omega f(x, \omega) \, dP(\omega) d\rho(x) = \rho(F)$$

where $F(x) = \int f(x, \omega) \, dP(\omega)$. Since $F \in C(X)$, by Remark 3.10 (iii) we get that $\rho \mapsto e\rho(f) = \rho(F)$ is continuous in the (non-random) narrow topology on $Pr(X)$ for every random continuous function f. The trace of the random narrow topology on $Pr(X)$ is the smallest topology on $Pr(X)$ making $\rho \mapsto e\rho(f)$ continuous for every $f \in C_\Omega(X)$, so the trace of the (random) narrow topology on $Pr(X)$ is weaker than the original (non-random) narrow topology on $Pr(X)$. Putting these observations together, we see that the narrow topology on $Pr(X)$ and the trace of the narrow topology on $Pr_\Omega(X)$ on $Pr(X)$ coincide.

This shows that $e : Pr(X) \to Pr_\Omega(X)$ is an *embedding* (i.e., continuous, injective, and open with respect to the trace topology on $e(Pr(X))$). We thus can treat the set $Pr(X)$ of non-random probability measures on X as a topological subspace of $Pr_\Omega(X)$. It is not hard to show that $Pr(X) \subset Pr_\Omega(X)$ is closed in the narrow topology of $Pr_\Omega(X)$. We will return to this question in a more general frame later. See Proposition 4.21 and Corollary 4.22.

Recall the canonical projection $\pi_X : X \times \Omega \to X$, $(x, \omega) \mapsto x$. Then π_X maps $Pr_\Omega(X)$ to $Pr(X)$ by

$$(\pi_X \mu)(B) = \mu(\pi_X^{-1}(B)) = \mu(B \times \Omega) = \int_\Omega \mu_\omega(B) \, dP(\omega) = E\mu(B) \quad (3.9)$$

for $B \in \mathscr{B}$; $\pi_X \mu$ is the marginal of μ on X, and E stands for expectation with respect to P. Instead of using the notation π_X we might as well use $\mu \mapsto E\mu$. Since we want to consider projection properties, we prefer π_X here. We need an elementary observation.

3.22 Lemma *For any $\mu \in Pr_\Omega(X)$ and for any $h : X \to \mathbb{R}$ measurable and integrable with respect to $\pi_X \mu$,*

$$\int_\Omega \int_X h(x) \, d\mu_\omega(x) \, dP(\omega) = \int_X h(x) \, d(\pi_X \mu)(x).$$

PROOF The assertion holds, by definition, for $h = 1_B$, $B \in \mathscr{B}$, and hence it holds for finite linear combinations of indicator functions. Approximating $h \geq 0$ by an increasing sequence of linear combinations of indicator functions, the assertion follows for h by monotone convergence, hence also for general $h = h^+ - h^-$.   □

As an immediate consequence we get for any $\mu \in Pr_\Omega(X)$ and $h : X \times \Omega \to \mathbb{R}$ measurable and integrable with respect to $(\pi_X \mu) \times P$, using the sloppy notation $Eh(x) = \int h(x, \omega) \, dP(\omega)$,

$$
\begin{aligned}
\int_\Omega \int_X h(x, \omega) \, d(\pi_X \mu)(x) \, dP(\omega) &= \int_X \int_\Omega h(x, \omega) \, dP(\omega) \, d(\pi_X \mu)(x) \\
&= \int_X Eh(x) \, d(\pi_X \mu)(x) \\
&= \int_\Omega \int_X Eh(x) \, d\mu_\omega(x) \, dP(\omega) \quad (3.10)
\end{aligned}
$$

(the first identity is Fubini, the second is the definition, and the third comes from Lemma 3.22). Following Remark 3.21 and interpreting $\pi_X \mu$ as a random measure, this translates into

$$\pi_X \mu(h) = \pi_X \mu(Eh) = \mu(Eh),$$

where the second term can be interpreted as integration on the (non-random) $Pr(X)$.

3.23 Remark Returning to $\pi_X : Pr_\Omega(X) \to Pr(X)$, Lemma 3.22 yields $(\pi_X\mu)(g) = \mu(g)$ for $g \in C(X)$ (where g is understood as a non-random element of $C_\Omega(X)$ on the right-hand side). Thus $\mu \mapsto (\pi_X\mu)(g)$ is continuous for any $g \in C(X)$. Consequently, π_X is continuous with respect to the narrow topologies on $Pr_\Omega(X)$ and $Pr(X)$, respectively. It is immediate that $\pi_X : Pr_\Omega(X) \to Pr(X)$ is surjective. The final topology induced on $Pr(X)$ by π_X is the collection of all $U \subset Pr(X)$ such that $\pi_X^{-1}U \subset Pr_\Omega(X)$ is open. Using the identification of $Pr(X)$ with the subspace of $Pr_\Omega(X)$ consisting of P-a.s. fixed measures, as described in Remark 3.21, π_X can be considered as a map from $Pr_\Omega(X)$ to itself. Then $\pi_X(Pr_\Omega(X)) = Pr(X)$, and $\pi_X^2 = \pi_X$. Proposition B.1 together with Corollary B.2 implies that the final topology induced by π_X on $Pr(X)$ coincides with the narrow topology on $Pr(X)$. Note that surjectivity of π_X implies that its final topology equals the quotient topology on $Pr(X)$, obtained by identifying two random measures if their images under π_X coincide.

Summarising Remarks 3.21 and 3.23, we get coincidence of the following three topologies on $Pr(X)$:

- the classical narrow topology

- the trace of the narrow topology of $Pr_\Omega(X)$ on $Pr(X) \subset Pr_\Omega(X)$ (which also coincides with the initial topology of the embedding $e : Pr(X) \to Pr_\Omega(X)$)

- the quotient topology induced by $\pi_X : Pr_\Omega(X) \to Pr(X)$.

We thus may imagine the space of random measures on X as a bundle over the space of deterministic measures on X with the projection π_X. The fibre with basepoint $\rho \in Pr(X)$ of this bundle is the set of all random measures μ with $\pi_X\mu = \rho$. This bundle is, however, neither linear nor finite-dimensional. It is not even trivialisable. The fibres have quite different structures. The fibre $\pi_X^{-1}(\delta_x)$ of the Dirac measure δ_x contains just one element (namely δ_x itself), whereas the fibre of a continuous (i.e., atomless) probability ρ contains infinitely many points.

Chapter 4

Prohorov Theory for Random Probability Measures

The notion of tightness in the context of non-random measures is well known. A set $G \subset Pr(X)$ is said to be *tight* if for every $\varepsilon > 0$ there exists a compact $K_\varepsilon \subset X$ such that $\rho(K_\varepsilon) \geq 1 - \varepsilon$ for all $\rho \in G$ (sometimes this is called 'uniformly tight'). One of the deepest results of the Prohorov theory is the relation between tightness and compactness stated in the following theorem. Since we restrict attention to the simplest case of a Polish state space X, we only state it for this situation.

4.1 Theorem (CLASSICAL PROHOROV THEOREM) *A set of probability measures $G \subset Pr(X)$ on a Polish space X has compact closure in the narrow topology if, and only if, it is tight.*

A proof of the Prohorov Theorem can be found in several places in the literature (see, for instance, Ethier and Kurtz [22], Theorem 2.2, pp. 104–105, or Parthasarathy [32], Theorem 6.7, p. 47).

Tightness for Random Measures

We will be concerned with a generalisation of the Prohorov theorem (Theorem 4.1) for random measures. The notion of tightness that we will use for random probability measures has been introduced by Valadier (see [37], p. 161).

4.2 Definition (TIGHTNESS FOR RANDOM MEASURES) A subset Γ of $Pr_\Omega(X)$ is said to be *tight*, if for every $\varepsilon > 0$ there exists a (non-random) compact set $C_\varepsilon \subset X$, such that $(\pi_X\gamma)(C_\varepsilon) \geq 1 - \varepsilon$ for every $\gamma \in \Gamma$, where $\pi_X\gamma$ is defined in (3.9).

This means that $\Gamma \subset Pr_\Omega(X)$ is tight if, and only if, its projection to the set of deterministic measures $\pi_X(\Gamma) \subset Pr(X)$ is tight. Therefore tightness of a set of random measures is a property which is characterised on the base space of the bundle $\pi_X : Pr_\Omega(X) \to Pr(X)$ already.

The following proposition gives criteria for tightness of a set Γ of random measures in terms of compact random sets. This covers the situation encountered in the theory of random dynamical systems, where one is interested in the set of all random measures supported by an invariant compact set. See Chapter 6, where these topics are discussed in some more detail.

4.3 Proposition *For $\Gamma \subset Pr_\Omega(X)$ consider the following assertions:*

(*i*) *For every $\varepsilon > 0$ there exists a compact random set $\omega \mapsto K_\varepsilon(\omega)$ such that, for every $\gamma \in \Gamma$,*

$$\gamma_\omega(K_\varepsilon(\omega)) \geq 1 - \varepsilon$$

for P-almost every ω.

(*ii*) *For every $\varepsilon > 0$ there exists a compact random set $\omega \mapsto K_\varepsilon(\omega)$ such that, for every $\gamma \in \Gamma$,*

$$\gamma(K_\varepsilon) = \int_\Omega \gamma_\omega(K_\varepsilon(\omega))\, dP(\omega) \geq 1 - \varepsilon.$$

(*iii*) Γ *is tight.*

Then (i) implies (ii), and (ii) and (iii) are equivalent.

PROOF Clearly (i) implies (ii), and (iii) implies (ii) by taking $K_\varepsilon(\omega) \equiv C_\varepsilon$ from Definition 4.2.

It remains to prove that (ii) implies (iii). For $\varepsilon > 0$ choose $C_\varepsilon \subset X$ compact, such that $P\{K_{\varepsilon/2}(\omega) \subset C_\varepsilon\} \geq 1 - \varepsilon/2$. Existence of C_ε with this property follows from Proposition 2.15. Then

$$\pi_X\gamma(C_\varepsilon) = \gamma(C_\varepsilon \times \Omega) = \int_\Omega \gamma_\omega(C_\varepsilon)\, dP(\omega)$$

$$\geq \int_{\{K_{\varepsilon/2}(\omega)\subset C_\varepsilon\}} \gamma_\omega(C_\varepsilon)\, dP$$

$$\geq \int_{\{K_{\varepsilon/2}(\omega) \subset C_\varepsilon\}} \gamma_\omega(K_{\varepsilon/2}(\omega))\, dP$$

$$\geq (1 - \varepsilon/2)^2 \geq 1 - \varepsilon$$

for every $\gamma \in \Gamma$.   □

Having introduced the notion of tightness for random measures, we turn to one of the main results, which is an almost complete restatement of the Prohorov theorem for random measures. A similar result is due to Valadier [37], Proposition 10 and Theorem 11, p. 162.

4.4 Theorem (PROHOROV THEOREM FOR RANDOM MEASURES) *Suppose that $\Gamma \subset Pr_\Omega(X)$. Then Γ is tight if, and only if, it is relatively compact with respect to the narrow topology. In this case it is also relatively sequentially compact (i.e., if $(\mu^n)_{n \in \mathbb{N}}$ is a sequence in Γ, then there exists a convergent subsequence $(\mu^{n_k})_{k \in \mathbb{N}}$).*

We will establish this result in the following, using the relations between $Pr(X)$ and $Pr_\Omega(X)$ considered in Remarks 3.21 and 3.23. Note that Theorem 4.4 contains the classical Prohorov theorem 4.1 (taking $(\Omega, \mathscr{F}, P)$ to be trivial, or alternatively, using the embedding $e : Pr(X) \to Pr_\Omega(X)$). We will approach the proof by taking the bundle point of view $\pi_X : Pr_\Omega(X) \to Pr(X)$. In particular, we definitely do not reprove the classical Prohorov theorem for the base $Pr(X)$, but rather start from it and pull it up the fibres. Furthermore, we avoid embedding X into a compact metrisable space. The complete proof will take some space, so the theorem is restated as Theorem 4.29, and the arguments yielding the proofs of the three assertions are listed.

The following lemma says that relative compactness implies tightness in the space of random measures. This is the immediate part of the generalisation of Theorem 4.4.

4.5 Lemma *Suppose that $\Gamma \subset Pr_\Omega(X)$ is relatively compact (in the narrow topology). Then Γ is tight.*

PROOF   Since $\pi_X : Pr_\Omega(X) \to Pr(X)$ is continuous (see Remark 3.23), relative compactness of Γ implies that $\pi_X(\Gamma)$ is relatively compact in $Pr(X)$, and thus tight by the Prohorov theorem 4.1.   □

To establish the other direction, i.e., to prove that tightness implies relative compactness, we need to investigate the narrow topology on $Pr_\Omega(X)$ closer. First we show that it is generated by random Lipschitz functions.

Random Lipschitz Functions

4.6 Definition The space of *random Lipschitz functions* is

$$\mathrm{BL}_\Omega(X) = \{g \in C_\Omega(X) : \omega \mapsto \|g(\cdot,\omega)\|_{\mathrm{BL}} \leq C \text{ for some } C \in \mathbb{R} \ P\text{-a.s.}\} \tag{4.1}$$

(see Definition 3.12 for $\|\cdot\|_{\mathrm{BL}}$).

4.7 Remark (i) Separability of X implies that $\omega \mapsto [g(\cdot,\omega)]_{\mathrm{L}}$ is measurable for $g \in C_\Omega(X)$ (see (3.4)). Therefore,

$$\omega \mapsto \|g(\cdot,\omega)\|_{\mathrm{BL}} = \max\Big\{[g(\cdot,\omega)]_{\mathrm{L}}, \sup_{x \in X}|g(x,\omega)|\Big\}$$

is measurable, and hence (4.1) makes sense.

(ii) The present notion of $\mathrm{BL}_\Omega(X)$ enforces a random Lipschitz function to have both its supremum and its Lipschitz constant to be bounded P-a.s., i.e., to be in $L^\infty(P)$. Thus the notion is not consistent with that of $C_\Omega(X)$, which only requires the supremum of a random continuous function to be in $L^1(P)$. Since it is the present notion we will use below, we refrain from introducing the collection of spaces obtained by combining the different requirements on supremum and Lipschitz constant to be in $L^1(P)$, $L^\infty(P)$ or other $L^p(P)$, $1 \leq p \leq \infty$. With the present notions we always have $\mathrm{BL}_\Omega(X) \subset C_\Omega(X)$, with any other of these spaces in between.

Exactly as for random continuous functions, any $g \in \mathrm{BL}(X)$ is understood as an element of $\mathrm{BL}_\Omega(X)$, and is addressed to as a *non-random* element of $\mathrm{BL}_\Omega(X)$. Also, any $h \in L^\infty(P)$ is understood as an element of $\mathrm{BL}_\Omega(X)$; compare the discussion following Definition 3.9.

(iii) The space $\mathrm{BL}_\Omega(X)$ is a vector space. It is also an algebra: For $f, g \in \mathrm{BL}_\Omega(X)$ Remark 3.13 (i) yields $\|f(\cdot,\omega)g(\cdot,\omega)\|_{\mathrm{BL}} \leq 2\|f(\cdot,\omega)\|_{\mathrm{BL}}\|g(\cdot,\omega)\|_{\mathrm{BL}}$ for all $\omega \in \Omega$. Furthermore,

$$\|\max\{f(\cdot,\omega), g(\cdot,\omega)\}\|_{\mathrm{BL}} \leq 2\max\Big\{\|f(\cdot,\omega)\|_{\mathrm{BL}}, \|g(\cdot,\omega)\|_{\mathrm{BL}}\Big\},$$

hence $f \vee g = \max\{f,g\} \in \mathrm{BL}_\Omega(X)$ by Remark 3.13 (iii). Thus $\mathrm{BL}_\Omega(X)$ is a vector lattice of real functions. Clearly $1 \in \mathrm{BL}_\Omega(X)$, where 1 denotes the constant function $(x,\omega) \mapsto 1$.

We note that $\mathrm{BL}_\Omega(X)$ containing constants implies that it is a *Stone* vector lattice. (The general definition of a Stone vector lattice requires that for any element f, $f \vee 1$ is also an element. If a vector lattice contains constants, then it is automatically a Stone vector lattice. See Dudley [19], Section 4.5, pp. 108–111.)

(iv) The smallest σ-algebra with respect to which all random Lipschitz functions are measurable is $\mathscr{B} \otimes \mathscr{F}$. In fact, measurability of $g \in \mathrm{BL}_\Omega(X)$ with respect to $\mathscr{B} \otimes \mathscr{F}$ follows from Lemma 1.1. On the other hand, for $B \in \mathscr{B}$ closed put $b_n(x) = 1 - \min\{nd(x, B), 1\}$, $n \in \mathbb{N}$. Then for $F \in \mathscr{F}$ the function $g_n(x, \omega) = b_n(x) 1_F(\omega)$ is a random Lipschitz function. Since g_n converges to $1_B 1_F$, $B \times F$ is in $\sigma(\mathrm{BL}_\Omega(X))$, the σ-algebra generated by $\mathrm{BL}_\Omega(X)$ on $X \times \Omega$, for every $B \subset X$ closed and $F \in \mathscr{F}$. Thus $\mathscr{B} \otimes \mathscr{F}$ is contained in $\sigma(\mathrm{BL}_\Omega(X))$.

Recall the definition of the δ-neighbourhood, $\delta > 0$, of $A \subset X$ arbitrary,

$$A^\delta = \{x \in X : d(x, A) < \delta\} = \{x : d(x, a) < \delta \text{ for some } a \in A\}.$$

4.8 Lemma *Suppose that $\omega \mapsto C(\omega)$ is a closed random set, and that $\mu \in Pr_\Omega(X)$ is a random measure. Then $\omega \mapsto \mu_\omega(C^\delta(\omega))$ converges to $\omega \mapsto \mu_\omega(C(\omega))$ for $\delta \to 0$, P-a.s. and in $L^1(P)$. In particular, for every $\varepsilon > 0$ there exists $g \in \mathrm{BL}_\Omega(X)$ with $1_{C(\omega)}(x) \le g(x, \omega) \le 1$ for all $(x, \omega) \in X \times \Omega$, such that*

$$\mu(g) = \int_\Omega \left(\int_X g(x, \omega) \, d\mu_\omega(x) \right) dP(\omega) \le \int_\Omega \mu_\omega(C(\omega)) \, dP(\omega) + \varepsilon.$$

PROOF Measurability of $\omega \mapsto \mu_\omega(C(\omega))$ and of $\omega \mapsto \mu_\omega(C^\delta(\omega))$ for $\delta > 0$ follows from Corollary 3.4 (invoking measurability of $\mathrm{graph}(C^\delta)$, which follows from Proposition 2.4). Closedness of C implies almost sure convergence of $\mu_\omega(C^\delta(\omega))$ to $\mu_\omega(C(\omega))$ for $\delta \to 0$. The convergence is, in addition, monotone, so $L^1(P)$-convergence follows from the monotone convergence theorem. Finally, for $\delta > 0$ put $g_\delta(x, \omega) = 1 - \min\{1, d(x, C(\omega))/\delta\}$. Then

$$1_{C(\omega)}(x) \le g_\delta(x, \omega) \le 1_{C^\delta(\omega)}(x), \tag{4.2}$$

and $[g_\delta(\cdot, \omega)]_{\mathrm{L}} = 1/\delta$. Integrating (4.2) with respect to μ and using $L^1(P)$-convergence of $\omega \mapsto \mu_\omega(C^\delta(\omega))$ to $\omega \mapsto \mu_\omega(C(\omega))$, we conclude that, for every δ sufficiently small, $g = g_\delta$ has the asserted properties. □

4.9 Proposition *If for some topology on $Pr_\Omega(X)$ (not necessarily the narrow topology) $\mu \mapsto \mu(g)$ is a continuous map from $Pr_\Omega(X)$ to $\mathbb{R}$ for every $g \in \mathrm{BL}_\Omega(X)$ with $0 \le g \le 1$ and $[g(\cdot, \omega)]_{\mathrm{L}} \le 1$ (P-a.s.), then $\mu \mapsto \mu(f)$ is continuous in this topology for every $f \in C_\Omega(X)$.*

PROOF For $g \in \mathrm{BL}_\Omega(X)$ with $0 \le g \le c$ and $[g(\cdot, \omega)]_{\mathrm{L}} \le c$ for some $0 < c \in \mathbb{R}$, continuity of $\mu \mapsto \mu(g)$ is immediate by multiplication with $1/c$.

In particular, this is true for g with $0 \leq g \leq 1$ and $[g(\cdot, \omega)]_L \leq c$ (P-a.s.) for some $c \in \mathbb{R}$.

It thus suffices to show that, for every $\mu \in Pr_\Omega(X)$, for every $f \in C_\Omega(X)$, and for every $\varepsilon > 0$, there exist $g_1, \ldots, g_N \in BL_\Omega(X)$ for some $N \in \mathbb{N}$ with $0 \leq g_k \leq 1$ and $[g_k(\cdot, \omega)]_L \leq c$, $1 \leq k \leq N$, for some $c \in \mathbb{R}$, and there exists $\delta > 0$ such that for every $\nu \in Pr_\Omega(X)$ with

$$\left| \int_{X \times \Omega} g_k d\mu - \int_{X \times \Omega} g_k d\nu \right| < \delta$$

for $0 \leq k \leq N$ it holds that

$$\left| \int_{X \times \Omega} f \, d\mu - \int_{X \times \Omega} f \, d\nu \right| < \varepsilon.$$

By Lemma 3.16 it suffices to prove this for $f \in C_\Omega(X)$ with $0 \leq f \leq 1$.

Suppose that $\mu \in Pr_\Omega(X)$, $\varepsilon > 0$, and suppose that $f \in C_\Omega(X)$ satisfies $0 \leq f \leq 1$. Choose $n \in \mathbb{N}$ with $n > 3/\varepsilon$. Put $C_k(\omega) = \{x \in X : f(x, \omega) \geq k/n\}$, $k = 0, \ldots, n$. Then C_k is a closed random set with respect to the universally completed σ-algebra of $\mathscr{F}$ by Remark 3.10 (i). Using first Lemma 4.8 and then Lemma 1.3 yields existence of $g_k \in BL_\Omega(X)$ with $1_{C_k(\omega)}(x) \leq g_k(x, \omega) \leq 1$, $k = 0, \ldots, n$, for P-almost all ω, such that

$$\mu(g_k) = \int_\Omega \left(\int_X g_k(x, \omega) \, d\mu_\omega(x) \right) dP(\omega) \leq \int_\Omega \mu_\omega(C_k(\omega)) \, dP(\omega) + \varepsilon/6 \quad (4.3)$$

for $k = 0, \ldots, n$. Suppose that $\nu \in Pr_\Omega(X)$ satisfies

$$\left| \int_{X \times \Omega} g_k d\mu - \int_{X \times \Omega} g_k d\nu \right| < \varepsilon/6 \quad (4.4)$$

for $1 \leq k \leq n$ (since $g_0 \equiv 1$ P-a.s. this is automatic for $k = 0$). From Lemma 1.4 we get, for any Borel probability measure γ on X and for every $\omega \in \Omega$,

$$\frac{1}{n} \sum_{k=1}^{n} \gamma(C_k(\omega)) \leq \int f(x, \omega) \, d\gamma(x)$$

$$\leq \frac{1}{n} \sum_{k=0}^{n} \gamma(C_k(\omega))$$

$$\leq \frac{1}{n} \sum_{k=0}^{n} \int_X g_k(x, \omega) \, d\gamma(x). \quad (4.5)$$

Applying (4.5) to ν_ω yields

$$\int_X f(x,\omega)\,d\nu_\omega(x) \;\leq\; \frac{1}{n}\sum_{k=0}^n \nu_\omega(C_k(\omega))$$

$$\leq\; \frac{1}{n}\sum_{k=0}^n \int_X g_k(x,\omega)\,d\nu_\omega(x) \qquad (4.6)$$

P-a.s., and applying (4.5) to μ_ω we get

$$\frac{1}{n}\sum_{k=1}^n \mu_\omega(C_k(\omega)) \leq \int_X f(x,\omega)\,d\mu_\omega(x) \qquad (4.7)$$

P-a.s. Consequently,

$$\int_\Omega \textstyle\int_X f(x,\omega)\,d\nu_\omega(x)\,dP(\omega)$$

$$\leq\; \frac{1}{n}\sum_{k=0}^n \int_\Omega \nu_\omega(C_k(\omega))\,dP(\omega)$$

$$\leq\; \frac{1}{n}\sum_{k=0}^n \int_\Omega \Big(\textstyle\int_X g_k(x,\omega)\,d\nu_\omega(x)\Big)dP(\omega) \qquad \text{by (4.6)}$$

$$<\; \frac{1}{n}\sum_{k=0}^n \Big(\int_\Omega \big(\textstyle\int_X g_k(x,\omega)\,d\mu_\omega(x)\big)dP(\omega) + \frac{\varepsilon}{6}\Big) \qquad \text{by (4.4)}$$

$$=\; \frac{1}{n}\sum_{k=0}^n \Big(\int_\Omega \big(\textstyle\int_X g_k(x,\omega)\,d\mu_\omega(x)\big)dP(\omega)\Big) + \frac{n+1}{6n}\varepsilon$$

$$\leq\; \frac{1}{n}\sum_{k=0}^n \Big(\int_\Omega \mu_\omega(C_k(\omega))\,dP(\omega) + \frac{\varepsilon}{6}\Big) + \frac{n+1}{6n}\varepsilon \qquad \text{by (4.3)}$$

$$=\; \frac{1}{n} + \frac{1}{n}\sum_{k=1}^n \Big(\int_\Omega \mu_\omega(C_k(\omega))\,dP(\omega)\Big) + \frac{n+1}{3n}\varepsilon$$

$$\leq\; \int_\Omega \textstyle\int_X f(x,\omega)\,d\mu_\omega(x)\,dP(\omega) + \Big(\frac{n+1}{n}+1\Big)\frac{\varepsilon}{3} \qquad \text{by (4.7)}$$

$$\leq\; \int_\Omega \textstyle\int_X f(x,\omega)\,d\mu_\omega(x)\,dP(\omega) + \varepsilon.$$

Applying the same considerations to $1-f$ instead of f we can define $g_{n+1},\dots,g_{2n} \in \mathrm{BL}_\Omega(X)$ such that

$$\left|\int_{X\times\Omega} g_k\,d\mu - \int_{X\times\Omega} g_k\,d\nu\right| < \varepsilon/6,$$

for $n + 1 \leq k \leq 2n$, implies

$$\int_\Omega \int_X (1 - f(x, \omega))\, d\nu_\omega(x)\, dP(\omega) < \int_\Omega \int_X (1 - f(x, \omega))\, d\mu_\omega(x)\, dP(\omega) + \varepsilon.$$

Consequently, for every $\nu \in Pr_\Omega(X)$ with

$$\left| \int_{X \times \Omega} g_k d\mu - \int_{X \times \Omega} g_k\, d\nu \right| < \varepsilon/6$$

for $1 \leq k \leq 2n$, we have

$$\left| \int_\Omega \left(\int_X f(x, \omega)\, d\mu_\omega(x) - \int_X f(x, \omega)\, d\nu_\omega(x) \right) dP(\omega) \right| < \varepsilon. \qquad \square$$

4.10 Corollary *The narrow topology on $Pr_\Omega(X)$ is already generated by the real-valued functions $\mu \mapsto \mu(g)$ on $Pr_\Omega(X)$, with $g \in \mathrm{BL}_\Omega(X)$ satisfying $0 \leq g \leq 1$ and $[g(\cdot, \omega)]_\mathrm{L} \leq 1$ P-a.s.*

Since

$$\left\{ g \in \mathrm{BL}_\Omega(X) : g \geq 0 \text{ and } \|g(\cdot, \omega)\|_{\mathrm{BL}} \leq 1 \text{ P-a.s.} \right\} \subset \mathrm{BL}_\Omega(X) \subset C_\Omega(X),$$

it follows, in particular, that the narrow topology on $Pr_\Omega(X)$ is generated by $\mathrm{BL}_\Omega(X)$, the Stone vector lattice of bounded random Lipschitz functions (compare Remark 4.7).

The Stone–Daniell Characterisation of Random Measures

We will use the Stone–Daniell Theorem (see Dudley [19], Theorem 4.5.2, p. 110) in order to obtain another characterisation of random measures. We state a version of the Stone–Daniell Theorem in a slightly reduced generality.

4.11 Theorem (STONE–DANIELL THEOREM) *Let V be a vector lattice of real valued functions on a set Z, containing constants. Denote by $\sigma(V)$ the σ-algebra generated by V on Z (the smallest σ-algebra such that every function from V is measurable). Suppose that $L : W \to \mathbb{R}$ is a pre-integral, which means*

- *L is linear, i.e., $L(f + \alpha g) = L(f) + \alpha L(g)$ for all $\alpha \in \mathbb{R}$ and $f, g \in V$*
- *L is nonnegative, i.e., $L(f) \geq 0$ for any $f \in V$ with $f \geq 0$*
- *$L(f_n) \downarrow 0$ for any sequence $(f_n)_{n \in \mathbb{N}}$ in V with $f_n \downarrow 0$.*

Then there is a measure μ on the measurable space $(Z, \sigma(V))$ such that $L(f) = \int f \, d\mu$ for every $f \in V$.

We are now in a position to prove that any linear nonnegative map from $\mathrm{BL}_\Omega(X)$ to $\mathbb{R}$ with marginals ρ on X and P on Ω is given by a random measure.

4.12 Proposition (STONE–DANIELL CHARACTERISATION OF RANDOM MEASURES) *Suppose that $L : \mathrm{BL}_\Omega(X) \to \mathbb{R}$ satisfies the following conditions:*

- *L is linear*
- *if $f \geq 0$ for $f \in \mathrm{BL}_\Omega(X)$ then $L(f) \geq 0$ (i.e., L is nonnegative)*
- *there exists $\rho \in Pr(X)$ such that $L(g) = \rho(g) = \int_X g \, d\rho$ for every non-random $g \in \mathrm{BL}(X)$*
- *$L(h) = E(h) = \int_\Omega h \, dP$ for every $h : \Omega \to \mathbb{R}$ essentially bounded and measurable.*

Then there exists a random measure $\mu \in Pr_\Omega(X)$ such that $L(f) = \mu(f)$ for all $f \in \mathrm{BL}_\Omega(X)$.

PROOF The space $\mathrm{BL}_\Omega(X)$ is a vector lattice containing constants, and $\sigma(\mathrm{BL}_\Omega(X)) = \mathscr{B} \otimes \mathscr{F}$ (see Remark 4.7 (iv)). We thus get existence of a probability measure μ on $(X \times \Omega, \mathscr{B} \otimes \mathscr{F})$ with $\pi_\Omega \mu = P$ (hence a random measure by Proposition 3.6) from the Stone–Daniell Theorem 4.11 once we have proved that L is a pre-integral. Having linearity and nonnegativity of L by assumption, it remains to prove that $L(f_n) \downarrow 0$ for any sequence $f_n \in \mathrm{BL}_\Omega(X)$, $n \in \mathbb{N}$, with $f_n(x, \omega) \downarrow 0$ for all (x, ω).

Given $\varepsilon > 0$, let $K \subset X$ be compact with $\rho(K) \geq 1 - \varepsilon$. Let $f_n \in \mathrm{BL}_\Omega(X)$, $n \in \mathbb{N}$, be a sequence with $f_n \downarrow 0$. Since $f_1 \in \mathrm{BL}_\Omega(X)$ there is $M \in \mathbb{R}$ such that $f_n \leq f_1 \leq M$ for all $n \in \mathbb{N}$. Furthermore, for each $n \in \mathbb{N}$ there is $C_n \in \mathbb{R}$ such that $\|f_n(\cdot, \omega)\|_{\mathrm{BL}} \leq C_n$ for P-almost all ω. Put $\delta_n = \varepsilon / C_n$ and $\chi_n(x) = 1 - \min\{1, \delta_n d(x, K)\}$. Then χ_n vanishes outside K^{δ_n}, the δ_n-neighbourhood of K. Furthermore, $(x, \omega) \mapsto f_n(x, \omega)\chi_n(x)$ defines an element of $\mathrm{BL}_\Omega(X)$ (cf. Remark 4.7 (iii)).

For all ω with $\|f_n(\cdot,\omega)\|_{\mathrm{BL}} \leq C_n$ we have for $x, z \in X$ arbitrary

$$
\begin{aligned}
&\left|f_n(x,\omega)\chi_n(x) - f_n(z,\omega)\chi_n(z)\right| \\
&= \left|\chi_n(x)\big(f_n(x,\omega) - f_n(z,\omega)\big) + f_n(z,\omega)\big(\chi_n(x) - \chi_n(z)\big)\right| \\
&\leq \chi_n(x)[f_n(\cdot,\omega)]_L\, d(x,z) + f_n(z,\omega) \\
&\leq C_n d(x,z) + f_n(z,\omega)
\end{aligned}
$$

(using $0 \leq \chi_n \leq 1$, hence $|\chi_n(x) - \chi_n(z)| \leq 1$). For $x \in K^{\delta_n}$ choose $z \in K$ with $d(x,z) < \delta_n$. Then

$$
\begin{aligned}
f_n(x,\omega)\chi_n(x) &\leq f_n(z,\omega)\chi_n(z) + |f_n(x,\omega)\chi_n(x) - f_n(z,\omega)\chi_n(z)| \\
&\leq f_n(z,\omega)\chi_n(z) + C_n d(x,z) + f_n(z,\omega) \\
&\leq 2\sup_{z\in K} f_n(z,\omega) + C_n\frac{\varepsilon}{C_n} \\
&= 2\sup_{z\in K} f_n(z,\omega) + \varepsilon.
\end{aligned}
$$

Since χ_n vanishes outside K^{δ_n} we can estimate

$$
\sup_{x\in X}\big(f_n(x,\omega)\chi_n(x)\big) = \sup_{x\in K^{\delta_n}}\big(f_n(x,\omega)\chi_n(x)\big) \leq 2\sup_{z\in K} f_n(z,\omega) + \varepsilon \tag{4.8}
$$

for all ω with $\|f_n(\cdot,\omega)\|_{\mathrm{BL}} \leq C_n$. Writing

$$
f_n(x,\omega) = f_n(x,\omega)\chi_n(x) + f_n(x,\omega)(1 - \chi_n(x))
$$

as sum of two functions from $\mathrm{BL}_\Omega(X)$, we can use linearity of L on $\mathrm{BL}_\Omega(X)$ to get

$$
L(f_n) = L(f_n\chi_n) + L(f_n(1 - \chi_n)). \tag{4.9}
$$

To estimate the second term recall that $f_n(1 - \chi_n) \leq M(1 - \chi_n)$, where $(1 - \chi_n) \in \mathrm{BL}_\Omega(X)$ is non-random. Nonnegativity of L implies

$$
L\big(f_n(1 - \chi_n)\big) \leq M\,L(1 - \chi_n) = M\int(1 - \chi_n)\,d\rho \leq M\rho(K^c) \leq M\varepsilon. \tag{4.10}
$$

Here we needed the assumption that L has marginal ρ on X.

Put $h_n(\omega) = \max_{z\in K} f_n(z,\omega)$, then $h_n \in L^\infty(P) \subset \mathrm{BL}_\Omega(X)$ (compare Remark 4.7 (ii)). Using nonnegativity of L together with (4.8) yields

$$
L(f_n\chi_n) \leq L(2h_n + \varepsilon) = 2E(h_n) + \varepsilon, \tag{4.11}
$$

using the assumption that L has marginal P on Ω. Continuity of f_n in x implies that $f_n(\cdot,\omega)|_K$ converges to 0 uniformly for each $\omega \in \Omega$, hence $h_n \downarrow 0$.

Applying monotone convergence to $h_n \downarrow 0$ on the probability space $(\Omega, \mathscr{F}, P)$, we get $E(h_n) \downarrow 0$.

Invoking (4.10) and (4.11) into (4.9) we obtain

$$L(f_n) \leq 2E(h_n) + \varepsilon + M\varepsilon,$$

and since $E(h_n)$ converges to 0 for $n \to \infty$ we may conclude

$$\limsup_{n \to \infty} L(f_n) \leq (1 + M)\varepsilon.$$

Now ε is arbitrary, thus we get $L(f_n) \downarrow 0$, so L is a pre-integral. $\qquad \square$

4.13 Remark The formulation of Proposition 4.12 refers to $BL_\Omega(X)$, and thus to the choice of a metric on X. But since $BL_\Omega(X) \subset C_\Omega(X)$ for any choice of the metric on X, any $L : C_\Omega(X) \to \mathbb{R}$ satisfying the four conditions of the proposition (linear, nonnegative, marginal $\rho \in Pr(X)$ on X, and marginal P on Ω) is given by a random measure. Proposition 4.12 thus gives another characterisation of random measures, independently of the choice of a metric on X.

A Criterion for Compactness in the Narrow Topology on Random Measures

The following theorem is one of the key results of the present approach. It combines Proposition 4.12 with the argument generally used to prove the Banach–Alaoglu theorem (which says essentially that the unit sphere of the dual of a linear space is compact in the weak* topology, see, e.g., Rudin [35], Theorem 3.15, pp. 66–68). We will use the fact that, for an arbitrary topological space Y and an arbitrary subset $Z \subset Y$, a set $C \subset Z$ is compact with respect to the trace topology on Z if and only if C is a compact subset of Y with respect to the original topology on Y. Note that $C' \subset Y$ compact with respect to the original topology does not imply $C' \cap Z$ compact with respect to the trace topology on Z. Compare Lemma B.3.

4.14 Theorem *If $G \subset Pr(X)$ is compact then $\pi_X^{-1}(G) \subset Pr_\Omega(X)$ is compact.*

PROOF For $f \in \mathrm{BL}_\Omega(X)$ put $M_f^+ = \sup\{\mu(f) : \mu \in Pr_\Omega(X)\}$ and $M_f^- = \inf\{\mu(f) : \mu \in Pr_\Omega(X)\}$, then $-C \leq M_f^- \leq M_f^+ \leq C$ for some $C \in \mathbb{R}$ (see (4.1)). Let

$$\mathfrak{C} = \prod_{f \in \mathrm{BL}_\Omega(X)} [M_f^-, M_f^+]$$

be the Cartesian product of the compact intervals $[M_f^-, M_f^+]$, $f \in \mathrm{BL}_\Omega(X)$, equipped with the product topology. Then $\mathfrak{C}$ is compact by the Tychonov theorem. Note that $\mathfrak{C}$ is the set of all functions $L : \mathrm{BL}_\Omega(X) \to \mathbb{R}$ with $L(f) \in [M_f^-, M_f^+]$ for all f (linear or not). A neighbourhood basis for the product topology on $\mathfrak{C}$ in $L_0 \in \mathfrak{C}$ is given by the sets $\{L \in \mathfrak{C} : |L(f_k) - L_0(f_k)| < \delta\}$ with $\delta > 0$, $n \in \mathbb{N}$ and $f_1, \ldots, f_n \in \mathrm{BL}_\Omega(X)$. Consider the map

$$\begin{aligned} I : Pr_\Omega(X) &\to \mathfrak{C}, \\ \mu &\mapsto \big(\mu(f)\big)_{f \in \mathrm{BL}_\Omega(X)}. \end{aligned}$$

Then I is one-to-one and continuous from $Pr_\Omega(X)$ to $\mathfrak{C}$. Furthermore, for $\nu \in Pr_\Omega(X)$ let

$$U = \big\{\mu \in Pr_\Omega(X) : |\mu(f_k) - \nu(f_k)| < \delta, \ k = 1, \ldots, n\big\} \qquad (4.12)$$

with $\delta > 0$ and $f_1, \ldots, f_n \in \mathrm{BL}_\Omega(X)$ be a neighbourhood of ν in the narrow topology. Then

$$I(U) = \big\{L \in \mathfrak{C} : |L(f_k) - \nu(f_k)| < \delta, \ k = 1, \ldots, n\big\} \cap I(Pr_\Omega(X)).$$

Since the collection of all $U \subset Pr_\Omega(X)$ of the form (4.12) with $\delta > 0$, $f_1, \ldots, f_n \in \mathrm{BL}_\Omega(X)$, $n \in \mathbb{N}$, and $\nu \in Pr_\Omega(X)$ varying form a basis of the narrow topology by Corollary 4.10, I maps open sets in $Pr_\Omega(X)$ (with respect to the narrow topology) to open sets in $I(Pr_\Omega(X))$ (with respect to the trace of the product topology of $\mathfrak{C}$ on $I(Pr_\Omega(X))$). Thus I is an open map onto its image $I(Pr_\Omega(X)) \subset \mathfrak{C}$ with respect to the trace topology, and hence I is a homeomorphism between $Pr_\Omega(X)$ and its image in $\mathfrak{C}$.

Now apply the same construction for $Pr(X)$ one level below. Put

$$C = \prod_{g \in \mathrm{BL}} (X)[M_g^-, M_g^+],$$

equipped with the product topology, and define

$$\begin{aligned} i : Pr(X) &\to C, \\ \rho &\mapsto \big(\rho(g)\big)_{g \in \mathrm{BL}(X)}. \end{aligned}$$

Also i is one-to-one, continuous, and open onto its image $i(Pr(X)) \subset C$.

Interpreting $\mathrm{BL}(X)$ as a subset of $\mathrm{BL}_\Omega(X)$, introduce the canonical projection $\pi : \mathfrak{C} \to C$. Clearly π is continuous and open.

From Lemma 3.22 we get $\pi \circ I = i \circ \pi_X$, which says that the following diagram commutes.

$$
\begin{array}{ccc}
Pr_\Omega(X) & \xrightarrow{\;\;I\;\;} & \prod\limits_{f \in \mathrm{BL}_\Omega(X)} [M_f^-, M_f^+] \\[2mm]
\Big\downarrow{\scriptstyle \pi_X} & & \Big\downarrow{\scriptstyle \pi} \\[4mm]
Pr(X) & \xrightarrow{\;\;i\;\;} & \prod\limits_{g \in \mathrm{BL}(X)} [M_g^-, M_g^+]
\end{array}
$$

For $D \subset C$ arbitrary this gives

$$
\pi_X^{-1}(i^{-1}(D)) = (i \circ \pi_X)^{-1}(D) = (\pi \circ I)^{-1}(D) = I^{-1}(\pi^{-1}(D)). \tag{4.13}
$$

Applying (4.13) for $D = i(M)$ with $M \subset Pr(X)$ arbitrary, and noting that $i^{-1}(i(M)) = M$ by injectivity of i, we get

$$
\pi_X^{-1}(M) = I^{-1}(\pi^{-1}(i(M))) \tag{4.14}
$$

for $M \subset Pr(X)$ arbitrary. Now let $G \subset Pr(X)$ be compact. Then $i(G) \subset C$ is compact, since i is continuous. Thus also $\pi^{-1}(i(G)) \subset \mathfrak{C}$ is compact in $\mathfrak{C}$ (as a closed subset of a compact set in a Hausdorff topology). Now

$$
\pi^{-1}(i(G)) = \big\{ L \in \mathfrak{C} : \text{there exists } \rho \in G \text{ such that } L(g) = \textstyle\int_X g\, d\rho
$$
$$
\text{for all (non-random) } g \in \mathrm{BL}(X) \big\}.
$$

By Proposition 4.12 for every linear and nonnegative $L \in \mathfrak{C}$ such that there exists $\rho \in Pr(X)$ with $L(g) = \int_X g\, d\rho$ for all (non-random) $g \in \mathrm{BL}(X)$, and $L(h) = \int_\Omega h\, dP$ for all $h \in L^\infty(P)$ (considered as a subspace of $\mathrm{BL}_\Omega(X)$), there exists $\mu \in Pr_\Omega(X)$ such that $I(\mu) = L$. Hence

$$
I(\pi_X^{-1}(G)) = \pi^{-1}(i(G))
$$
$$
\cap \big\{ L \in \mathfrak{C} : L(f + \alpha g) = L(f) + \alpha L(g)
$$
$$
\text{for all } \alpha \in \mathbb{R},\ f, g \in \mathrm{BL}_\Omega(X) \big\}
$$
$$
\cap \big\{ L \in \mathfrak{C} : L(f) \geq 0 \text{ for all } f \geq 0 \big\}
$$
$$
\cap \big\{ L \in \mathfrak{C} : L(h) = \textstyle\int_\Omega h\, dP
$$
$$
\text{for all } h \in L^\infty(P)\ (\subset \mathrm{BL}_\Omega(X)) \big\}.
$$

Each of these four sets is compact in $\mathfrak{C}$. Since their intersection $I(\pi_X^{-1}(G))$ is a subset of $I(Pr_\Omega(X))$, $I(\pi_X^{-1}(G))$ is also compact in the relative topology of $I(Pr_\Omega(X))$ by Lemma B.3. Finally, since I is a homeomorphism between $Pr_\Omega(X)$ and $I(Pr_\Omega(X))$, $\pi_X^{-1}(G)$ is compact as well.   □

4.15 Corollary *Suppose that $\Gamma \subset Pr_\Omega(X)$ is tight. Then Γ is relatively compact.*

PROOF We have to prove that the closure $\mathrm{cl}(\Gamma)$ of Γ is compact. Continuity of π_X on $Pr_\Omega(X)$ implies $\pi_X(\mathrm{cl}(\Gamma)) \subset \mathrm{cl}(\pi_X(\Gamma))$. Now tightness of $\Gamma \subset Pr_\Omega(X)$ just means (by Definition 4.2) that $\pi_X(\Gamma) \subset Pr(X)$ is tight. Thus $\mathrm{cl}(\pi_X(\Gamma))$ is compact by Theorem 4.1 (the classical Prohorov theorem). Theorem 4.14 yields $\pi_X^{-1}(\mathrm{cl}(\pi_X(\Gamma)))$ compact. Putting all this together gives

$$\mathrm{cl}(\Gamma) \subset \pi_X^{-1}(\pi_X(\mathrm{cl}(\Gamma))) \subset \pi_X^{-1}(\mathrm{cl}(\pi_X(\Gamma))),$$

whence $\mathrm{cl}(\Gamma)$ is a closed subset of a compact set, and hence compact itself.   □

Taking together Lemma 4.5 and Corollary 4.15, we have established equivalence of (i) and (ii) of Theorem 4.4.

Metrisability of the Narrow Topology

We want to prove that tightness implies sequential compactness. However, in contrast to the classical narrow topology on non-random measures, the narrow topology of $Pr_\Omega(X)$ is not metrisable in general. It will be shown now that it is metrisable as soon as the probability space is countably generated (mod P). Given a probability space $(\Omega, \mathscr{F}, P)$, the σ-algebra $\mathscr{F}$ is said to be *countably generated* (mod P) if there is a countable algebra $\mathscr{F}_0 \subset \mathscr{F}$ such that for every $F \in \mathscr{F}$ there exists G in the σ-algebra $\sigma(\mathscr{F}_0)$ generated by $\mathscr{F}_0$, such that $P(F \bigtriangleup G) = 0$.

4.16 Theorem *Suppose that $(\Omega, \mathscr{F}, P)$ is countably generated (mod P), and let $\mathscr{F}_0 = \{G_m : m \in \mathbb{N}\}$ be a countable algebra generating $\mathscr{F}$ (mod P) with $\Omega = G_0$. For $\mu, \nu \in Pr_\Omega(X)$ put*

$$d(\mu, \nu) = \sum_{m \in \mathbb{N}} \frac{1}{2^m} \sup\left\{ \int_{G_m} (\mu_\omega(g) - \nu_\omega(g))\, dP(\omega) : g \in \mathrm{BL}(X), \right.$$
$$\left. 0 \le g \le 1,\ [g]_{\mathrm{L}} \le 1 \right\}.$$

Then d is a metric on $Pr_\Omega(X)$, which metrises the narrow topology. Furthermore, for any $g \in BL_\Omega(X)$ with $0 \le g \le 1$ and $[g(\cdot, \omega)]_L \le 1$ for P-almost all ω

$$|\mu(g) - \nu(g)| \le d(\mu, \nu) + 2\sqrt{5}\, d(\mu, \nu)^{1/2} \tag{4.15}$$

for all $\mu, \nu \in Pr_\Omega(X)$.

PROOF The triangle inequality for d is immediate. If $g : X \to \mathbb{R}$ satisfies $g \in BL(X)$, $0 \le g \le 1$ and $[g]_L \le 1$ then also $1 - g : X \to \mathbb{R}$ has these properties, and hence d is symmetric and nonnegative. Furthermore, $d(\mu, \nu) = 0$ implies $\mu = \nu$ by Lemma 3.14, so d is a metric. Due to $\Omega = G_0$ we have

$$\int_\Omega \left(\mu_\omega(g) - \nu_\omega(g) \right) dP(\omega) \le d(\mu, \nu) \tag{4.16}$$

for every $g \in BL(X)$ with $0 \le g \le 1$ and $[g]_L \le 1$. Note that for any such g and $\mu, \nu \in Pr_\Omega(X)$, $\mu_\omega(g) \le 1 \le 1 + \nu_\omega(g)$ P-a.s., hence $\mu_\omega(g) - \nu_\omega(g) \le 1$ P-a.s., which implies $d(\mu, \nu) \le 2$.

A remark on the notation: In the following integrals over the three spaces X, Ω, and $X \times \Omega$ appear. Usually the set or space to which the corresponding integration refers to is denoted below the integral sign, except for integration over $X \times \Omega$.

STEP 1 Pick $\mu \in Pr_\Omega(X)$ and $0 < \delta \le 1$. The projection $\pi_X \mu$ is a Borel probability measure on X, hence for every $\delta > 0$ there exists a compact $K \subset X$ with $\pi_X \mu(K) \ge 1 - \delta$. Put $B = \{h \in BL(X) : \|h\|_{BL} \le 1\}$. Then the restriction $B|_K = \{h|_K : h \in B\}$ is a compact subset of $C(K)$ with respect to the supremum topology by the Arzelà–Ascoli theorem (see, e.g., Dudley [19], Theorem 2.4.7, p. 40). So for any $\eta > 0$ there exist $g_1, \ldots, g_N \in B$ such that for every $h \in B$

$$\min_{1 \le n \le N} \left(\sup_{z \in K} |h(z) - g_n(z)| \right) < \eta \tag{4.17}$$

(which means that for every h at least one of the g_n is η-close to h on K). Note that $\sup_{z \in K} |h(z) - g_n(z)| < \eta$ implies

$$\sup_{z \in K^\delta} |h(z) - g_n(z)| < \eta + 2\delta \tag{4.18}$$

(for any $g_n, h \in B$), since for $z \in K$ and $d(x, z) < \delta$

$$\begin{aligned}
|h(x) - g_n(x)| &\le |h(x) - h(z)| + |h(z) - g_n(z)| + |g_n(z) - g_n(x)| \\
&\le [h]_L\, d(x, z) + \eta + [g_n]_L\, d(x, z) \\
&< \eta + 2\delta.
\end{aligned}$$

Clearly $\sup_{x \in X} |h(x) - g_n(x)| \le 2$ for $1 \le n \le N$.

Finally, let h_0 be a bounded Lipschitz function on X with $1_K \leq h_0 \leq 1_{K^\delta}$ and $[h_0]_L \leq 1/\delta$ (put, e.g., $h_0(x) = \max\{0, 1 - d(x, K)/\delta\}$). Then $g_0 = \delta h_0 \in BL(X)$ with $0 \leq g_0 \leq 1$ and $[g_0]_L \leq 1$. Consequently, for $\nu \in Pr_\Omega(X)$ arbitrary,

$$
\begin{aligned}
\int_\Omega \nu_\omega(K^\delta) \, dP(\omega) &\geq \int_\Omega \nu_\omega(h_0) \, dP(\omega) \\
&= \int_\Omega \mu_\omega(h_0) \, dP(\omega) - \int_\Omega (\mu_\omega(h_0) - \nu_\omega(h_0)) \, dP(\omega) \\
&\geq \int_\Omega \mu_\omega(K) \, dP(\omega) - \frac{1}{\delta} \int_\Omega (\mu_\omega(g_0) - \nu_\omega(g_0)) \, dP(\omega) \\
&\geq (1 - \delta) - \frac{1}{\delta} d(\mu, \nu),
\end{aligned}
$$

using (4.16) for the last inequality. This yields

$$
\int_\Omega \nu_\omega(K^{\delta c}) \, dP(\omega) < \delta + \frac{1}{\delta} d(\mu, \nu) \tag{4.19}
$$

for any $\nu \in Pr_\Omega(X)$, where $K^{\delta c} = (K^\delta)^c$.

Note that the above considerations are concerned with properties of μ and the associated K only (and, of course, with δ and η). All the functions g_n as well as h_0 are non-random, not depending on ω. Also note that (4.19) uses (4.16), which in turn invokes the assumption $G_0 = \Omega$. Coefficients have to be modified if this assumption is dropped.

STEP 2   To conclude that the metric topology induced by d is finer than the narrow topology, it suffices to prove that for every $\mu \in Pr_\Omega(X)$, every $f \in C_\Omega(X)$, and every $\varepsilon > 0$ there exists $\delta > 0$ such that $d(\mu, \nu) < \delta$ implies $|\mu(f) - \nu(f)| < \varepsilon$. By Corollary 4.10 it suffices to prove this for $f \in BL_\Omega(X)$ with $0 \leq f \leq 1$ and $[f(\cdot, \omega)]_L \leq 1$ for P-almost all ω. The assertion thus follows once we have established (4.15), which says

$$
|\mu(f) - \nu(f)| \leq d(\mu, \nu) + 2\sqrt{5} d(\mu, \nu)^{1/2}
$$

for every such f. Pick $\mu \in Pr_\Omega(X)$, $0 < \delta \leq 1$, $\eta > 0$ and choose $K \subset X$ compact with $\pi_X \mu(K) \geq 1 - \delta$. Then (4.19) holds for any $\nu \in Pr_\Omega(X)$. Next, using the considerations of Step 1, choose $N \in \mathbb{N}$ and $g_1, \ldots, g_N \in B$ such that (4.17) (and consequently (4.18)) holds. Then pick $f \in BL_\Omega(X)$ with $0 \leq f \leq 1$ and $[f(\cdot, \omega)]_L \leq 1$ for all $\omega \in \Omega$, clearly $f(\cdot, \omega) \in B$ for all $\omega \in \Omega$. For $n = 1, \ldots, N$ put

$$
F_n' = \{\omega \in \Omega : \sup_{x \in K} |f(x, \omega) - g_n(x)| < \eta\}
$$

with $g_n \in B$ chosen above. Then $\Omega = \bigcup_{n=1}^{N} F'_n$ by (4.17), and for $\omega \in F'_n$

$$\sup_{x \in K} |f(x, \omega) - g_n(x)| < \eta. \qquad (4.20)$$

Put $F_1 = F'_1$ and

$$F_n = F'_n \backslash \left(\bigcup_{i=1}^{n-1} F'_i \right)$$

for $2 \leq n \leq N$ to obtain a finite partition $\{F_1, \ldots, F_n\}$ of Ω into measurable sets. Since $\mathscr{F}_0 = \{G_m : m \in \mathbb{N}\}$ generates $\mathscr{F}$ (mod P), for every $\eta > 0$ there exist $m(n) \in \mathbb{N}$ such that

$$P(F_n \triangle G_{m(n)}) < \frac{\eta}{2N} \qquad (4.21)$$

for $n = 1, \ldots, N$ (see Gänssler and Stute [24], Satz 1.4.14, p. 29). Now, for $\nu \in Pr_\Omega(X)$ arbitrary,

$$\begin{aligned}
\mu(f) - \nu(f) &= \int_\Omega \left(\int_X f(x, \omega)\, d\mu_\omega(x) - \int_X f(x, \omega)\, d\nu_\omega(x) \right) dP(\omega) \\
&= \sum_{n=1}^{N} \int_{F_n} \left(\int_K \big(f(x, \omega) - g_n(x)\big)\, d\mu_\omega(x) \right. \\
&\qquad + \int_{K^c} \big(f(x, \omega) - g_n(x)\big)\, d\mu_\omega(x) \\
&\qquad + \int_{K^\delta} \big(f(x, \omega) - g_n(x)\big)\, d\nu_\omega(x) \\
&\qquad + \int_{K^{\delta c}} \big(f(x, \omega) - g_n(x)\big)\, d\nu_\omega(x) \\
&\qquad \left. + \int_X \mu_\omega(g_n) - \nu_\omega(g_n) \right) dP(\omega). \qquad (4.22)
\end{aligned}$$

We estimate the five terms of the right-hand side of (4.22) individually. Invoking (4.20), we have for $\omega \in F_n \subset F'_n$

$$\int_K \big(f(x, \omega) - g_n(x)\big)\, d\mu_\omega(x) \leq \eta \mu_\omega(K) \leq \eta,$$

hence

$$\sum_{n=1}^{N} \int_{F_n} \left(\int_K \big(f(x, \omega) - g_n(x)\big)\, d\mu_\omega(x) \right) dP(\omega) \leq \eta \sum_{n=1}^{N} P(F_n) = \eta. \qquad (4.23)$$

Similarly, we get with (4.18) for $\omega \in F_n$

$$\int_{K^\delta} (f(x,\omega) - g_n(x)) \, d\nu_\omega(x) \leq (\eta + 2)\delta\nu(K^\delta) \leq \eta + 2\delta,$$

hence

$$\sum_{n=1}^{N} \int_{F_n} \left(\int_{K^\delta} (f(x,\omega) - g_n(x)) \, d\nu_\omega(x) \right) dP(\omega) \leq \eta + 2\delta. \tag{4.24}$$

Next, using $f(x,\omega) - g_n(x) \leq 2$ and $\pi_X(K) = \int_\Omega \mu_\omega(K) \, dP(\omega) \leq \delta$, we get

$$\begin{aligned}
\sum_{n=1}^{N} \int_{F_n} \left(\int_{K^c} (f(x,\omega) - g_n(x)) \, d\mu_\omega(x) \right) dP(\omega) &\leq \sum_{n=1}^{N} \int_{F_n} 2\mu_\omega(K^c) \, dP(\omega) \\
&= 2\pi_X\mu(K) \\
&\leq 2\delta. \tag{4.25}
\end{aligned}$$

From (4.19) we obtain

$$\begin{aligned}
\sum_{n=1}^{N} \int_{F_n} \left(\int_{K^{\delta c}} (f(x,\omega) - g_n(x)) \, d\nu_\omega(x) \right) dP(\omega) &\leq \sum_{n=1}^{N} \int_{F_n} 2\nu_\omega(K^{\delta c}) \, dP(\omega) \\
&\leq \delta + \frac{1}{\delta} d(\mu,\nu). \tag{4.26}
\end{aligned}$$

Finally, using

$$\begin{aligned}
\int_D h \, dP &= \int_E h \, dP + \int_{D \setminus E} h \, dP - \int_{E \setminus D} h \, dP \\
&\leq \int_E h \, dP + \int_{D \triangle E} \sup |h| \, dP
\end{aligned}$$

for arbitrary measurable sets D, E, and for h integrable, we get

$$\begin{aligned}
&\int_{F_n} (\mu_\omega(g_n) - \nu_\omega(g_n)) \, dP(\omega) \\
&\leq \int_{G_{m(n)}} (\mu_\omega(g_n) - \nu_\omega(g_n)) \, dP(\omega) + 2P(F_n \triangle G_{m(n)}).
\end{aligned}$$

Invoking (4.21), this gives

$$\sum_{n=1}^{N} \int_{F_n} (\mu_\omega(g_n) - \nu_\omega(g_n)) \, dP(\omega)$$

$$\leq \sum_{n=1}^{N} \int_{G_{m(n)}} (\mu_\omega(g_n) - \nu_\omega(g_n)) dP(\omega) + 2 \sum_{n=1}^{N} P(G_{m(n)} \triangle F_n)$$

$$\leq \sum_{n=1}^{N} \sup \left\{ \int_{G_{m(n)}} (\mu_\omega(g) - \nu_\omega(g)) dP(\omega) : g \in BL(X), \right.$$
$$\left. 0 \leq g \leq 1, \ [g]_L \leq 1 \right\}$$

$$+ 2N \frac{\eta}{2N}$$

$$\leq d(\mu, \nu) + \eta. \tag{4.27}$$

Estimating the right-hand side of the identity (4.22) using (4.23), (4.25), (4.24), (4.26), and (4.27), we obtain

$$\mu(f) - \nu(f) \leq \eta + 2\delta + \eta + 2\delta + \delta + \frac{1}{\delta} d(\mu, \nu) + d(\mu, \nu) + \eta$$

$$= 5\delta + \frac{1}{\delta} d(\mu, \nu) + d(\mu, \nu) + 3\eta.$$

This holds, for $0 < \delta \leq 1$ fixed, for $\eta > 0$ arbitrary; hence we get

$$\mu(f) - \nu(f) \leq 5\delta + \frac{1}{\delta} d(\mu, \nu) + d(\mu, \nu)$$

for every $0 < \delta \leq 1$. Since $d(\mu, \nu) \leq 2$, we have

$$\inf_{0 < \delta \leq 1} \left(5\delta + \frac{d(\mu, \nu)}{\delta} \right) = 2\sqrt{5 d(\mu, \nu)},$$

so we obtain

$$\mu(f) - \nu(f) \leq d(\mu, \nu) + 2\sqrt{5} \, d(\mu, \nu)^{1/2}$$

for every $f \in BL_\Omega(X)$ with $0 \leq f \leq 1$ and $[f(\cdot, \omega)]_L \leq 1$ P-a.s. For every such f also $1 - f$ has these properties, so we can apply the same argument to $1 - f$ to obtain

$$\nu(f) - \mu(f) \leq d(\mu, \nu) + 2\sqrt{5} \, d(\mu, \nu)^{1/2}.$$

This implies (4.15), and hence the assertion that the metric topology induced by d is finer than the narrow topology.

STEP 3 To prove that the narrow topology is finer than the metric topology induced by d, it suffices to show that for every $\mu \in Pr_\Omega(X)$ and every $\varepsilon > 0$ there exist $f_1, \ldots, f_J \in C_\Omega(X)$ for some $J \in \mathbb{N}$ and $\delta > 0$ such that every $\nu \in Pr_\Omega(X)$ with

$$|\mu(f) - \nu(f)| = \left| \int f_n \, d\mu - \int f_n \, d\nu \right| < \delta$$

for $1 \leq n \leq N$ it holds that
$$d(\mu, \nu) < \varepsilon.$$

In order to prove this, let $\mu \in Pr_\Omega(X)$ and $\varepsilon > 0$ be given, and choose $M \in \mathbb{N}$ with $2^M \geq 4/\varepsilon$. Put $\delta = \varepsilon/45$. As in Step 1 let $K \subset X$ be a compact set such that $\pi_X \mu(K) \geq 1 - \delta$. Choose $g_1, \ldots, g_N \in B = \{h \in \mathrm{BL}(X) : \|h\|_{\mathrm{BL}} \leq 1\}$ such that for every $h \in B$

$$\sup_{z \in K} |h(z) - g_n(z)| < \delta$$

and put $g_0(x) = \max\{0, 1 - d(x, K)/\delta\}$. Repeating an argument of Step 1 in a slightly modified version we get for every $\nu \in Pr_\Omega(X)$

$$\begin{aligned}
\int_\Omega \nu_\omega(K^\delta)\, dP(\omega) \;&\geq\; \int_\Omega \nu_\omega(g_0)\, dP(\omega) \\
&=\; \int_\Omega \mu_\omega(g_0)\, dP(\omega) - \int_\Omega \big(\mu_\omega(g_0) - \nu_\omega(g_0)\big)\, dP(\omega) \\
&\geq\; \int_\Omega \mu_\omega(K)\, dP(\omega) - \int_\Omega \big(\mu_\omega(g_0) - \nu_\omega(g_0)\big)\, dP(\omega),
\end{aligned}$$

hence

$$\int_\Omega \nu_\omega(K^{\delta c})\, dP(\omega) < \delta + \int_\Omega \big(\mu_\omega(g_0) - \nu_\omega(g_0)\big)\, dP(\omega), \qquad (4.28)$$

where $K^{\delta c} = (K^\delta)^c$.

Define $f_{nm} \in C_\Omega(X)$ by

$$f_{nm}(x, \omega) = g_n(x) 1_{G_m}(\omega)$$

for $n = 0, \ldots, N$ and $m = 0, \ldots, M$. Then (4.28) implies

$$\int_\Omega \nu_\omega(K^{\delta c})\, dP(\omega) < \delta + \left| \int f_{00}\, d\mu - \int f_{00}\, d\nu \right| \qquad (4.29)$$

(using $G_0 = \Omega$). Now suppose that $\nu \in Pr_\Omega(X)$ satisfies

$$|\mu(f_{nm}) - \nu(f_{nm})| = \left| \int f_{nm}\, d\mu - \int f_{nm}\, d\nu \right| < \delta = \frac{\varepsilon}{45}$$

for all $n = 0, \ldots, N$ and $m = 0, \ldots, M$. We want to prove that this implies $d(\mu, \nu) < \varepsilon$.

To achieve this, consider $g \in BL(X)$ with $0 \le g \le 1$ and $[g]_L \le 1$. For any such g there exists n, $1 \le n \le N$, such that the following three conditions are satisfied:

$$\sup_{z \in K} |g(z) - g_n(z)| < \delta,$$
$$\sup_{z \in K^\delta} |g(z) - g_n(z)| < 3\delta, \tag{4.30}$$
$$\sup_{x \in X} |g(x) - g_n(x)| \le 2$$

(see (4.17) and (4.18) from Step 1). For any m, $0 \le m \le M$, then

$$\int_{G_m} \big(\mu_\omega(g) - \nu_\omega(g)\big) dP(\omega)$$

$$= \int_\Omega \Big[\int_X g(x) 1_{G_m}(\omega) d\mu_\omega(x) - \int_X g(x) 1_{G_m}(\omega) d\nu_\omega(x)\Big] dP(\omega)$$

$$= \int_\Omega \Big[\int_X (g(x) - g_n(x)) 1_{G_m}(\omega) d\mu_\omega(x)$$
$$- \int_X (g(x) - g_n(x)) 1_{G_m}(\omega) d\nu_\omega(x)\Big] dP(\omega)$$

$$+ \int_\Omega \Big[\int_X g_n(x) 1_{G_m}(\omega) d\mu_\omega(x) - \int_X g_n(x) 1_{G_m}(\omega) d\nu_\omega(x)\Big] dP(\omega)$$

$$= \int_{G_m} \Big[\int_K (g(x) - g_n(x)) d\mu_\omega(x)\Big] dP(\omega)$$

$$+ \int_{G_m} \Big[\int_{K^c} (g(x) - g_n(x)) d\mu_\omega(x)\Big] dP(\omega)$$

$$+ \int_{G_m} \Big[\int_{K^\delta} (g_n(x) - g(x)) d\nu_\omega(x)\Big] dP(\omega)$$

$$+ \int_{G_m} \Big[\int_{K^{\delta c}} (g_n(x) - g(x)) d\nu_\omega(x)\Big] dP(\omega)$$

$$+ \int_{X \times \Omega} f_{nm}(x, \omega) d\mu(x, \omega) - \int_{X \times \Omega} f_{nm}(x, \omega) d\nu(x, \omega).$$

Using (4.30) to estimate the first three of the five terms above we obtain first

$$\int_{G_m} \int_K (g(x) - g_n(x)) d\mu_\omega(x) dP(\omega) \le \delta \int_{G_m} \mu_\omega(K) dP(\omega) \le \delta.$$

Next, since $\pi_X \mu(K^c) \le \delta$,

$$\int_{G_m} \int_{K^c} (g(x) - g_n(x)) d\mu_\omega(x) dP(\omega) \le 2 \int_{G_m} \mu_\omega(K^c) dP(\omega)$$

$$\le 2 \int_\Omega \mu_\omega(K^c) dP(\omega) \le 2\delta$$

and, finally,

$$\int_{G_m} \int_{K^\delta} \big(g_n(x) - g(x)\big) \, d\nu_\omega(x) \, dP(\omega) \le 3\delta \int_{G_m} \nu_\omega(K^\delta) \, dP(\omega) \le 3\delta.$$

To estimate the fourth term we invoke (4.29) to get

$$\int_{G_m} \int_{K^{\delta c}} \big(g_n(x) - g(x)\big) \, d\nu_\omega(x) \, dP(\omega)$$

$$\le 2 \int_{G_m} \nu_\omega(K^{\delta c}) \, dP(\omega) \le 2 \int_\Omega \nu_\omega(K^{\delta c}) \, dP(\omega)$$

$$< 2\Big(\delta + \Big|\int f_{00} \, d\mu - \int f_{00} \, d\nu\Big|\Big) \le 4\delta.$$

Consequently,

$$\int_{G_m} \big(\mu_\omega(g) - \nu_\omega(g)\big) dP(\omega)$$

$$\le (\delta + 2\delta + 3\delta + 4\delta) + \Big|\int f_{nm}(x, \omega) \, d\mu(x, \omega) - \int f_{nm}(x, \omega) \, d\nu(x, \omega)\Big|$$

$$< 11\delta.$$

This holds for every $g \in \mathrm{BL}(X)$ with $0 \le g \le 1$ and $[g]_\mathrm{L} \le 1$, and for every $0 \le m \le M$. We therefore get, introducing the numbers R_m to abbreviate notation,

$$R_m = \sup\Big\{\int_{G_m} \big(\mu_\omega(g) - \nu_\omega(g)\big) dP(\omega) : 0 \le g \le 1, \ [g]_\mathrm{L} \le 1\Big\}$$

$$\le \begin{cases} 11\delta & \text{for } 0 \le m \le M, \\ 2 & \text{for } m > M. \end{cases}$$

Putting everything together we estimate

$$d(\mu, \nu) = \sum_{m \in \mathbb{N}} \frac{1}{2^m} R_m$$

$$= \sum_{m \le M} \frac{R_m}{2^m} + \sum_{m > M} \frac{R_m}{2^m}$$

$$\le 11\delta \sum_{m \in \mathbb{N}} \frac{1}{2^m} + \sum_{m \ge M} \frac{1}{2^m}$$

$$= 22\delta + \frac{2}{2^M}$$

$$\le \frac{22}{45}\varepsilon + \frac{1}{2}\varepsilon < \varepsilon$$

by the choice of M and δ, proving the assertion. □

4.17 Remark (i) The assumption $\Omega = G_0$ of Theorem 4.16 has been made only for convenience. The result remains true without this assumption. However, when dropping this assumption, estimate (4.15) does not hold as it stands. The coefficients of the right-hand side have to be modified then, depending on m with $\Omega = G_m \pmod{P}$.

(ii) To denote the metric on $Pr_\Omega(X)$ we used d again, so d denotes metrics on X, $Pr(X)$ and $Pr_\Omega(X)$ now. Since X is embedded into $Pr(X)$ by $x \mapsto \delta_x$, and $Pr(X)$ is embedded into $Pr_\Omega(X)$ (see Remark 3.21), the $Pr_\Omega(X)$ metric d restricted to $Pr(X)$ topologises the narrow topology on $Pr(X)$, and the $Pr(X)$ metric restricted to X topologises the topology of X. Thus, using the same letter to denote metrics on all those three spaces is not completely unjustified. For $\rho, \zeta \in Pr(X)$ we have

$$d(\rho, \zeta) = \left(\sum_{m \in \mathbb{N}} \frac{P(G_m)}{2^m} \right) \sup\left\{ \rho(g) - \zeta(g) : g \in \mathrm{BL}(X),\ 0 \le g \le 1,\ [g]_\mathrm{L} \le 1 \right\},$$

$$(4.31)$$

so the restriction of the metric d to the non-random measures $Pr(X)$ is a metric equivalent to the dual-bounded Lipschitz metric

$$\beta(\rho, \zeta) = \sup\left\{ |\rho(g) - \zeta(g)| : g \in \mathrm{BL}(X),\ [g]_\mathrm{L} + \sup_{x \in X} |g(x)| \le 1 \right\} \quad (4.32)$$

(see Dudley [19], Proposition 11.3.2, p. 310).

Note that the constants in the estimates yielding equivalence of the two metrics involve $\sum P(G_m)/2^m$, so they depend on the choice of the enumeration of the G_m, not just on $\Omega = G_0$. But still $\Omega = G_0$ implies $\sum P(G_m)/2^m \ge 1$, whereas without any restriction on the position of Ω in the enumeration, $\sum P(G_m)/2^m$ can be made arbitrarily small.

(iii) The projection $\pi_X : Pr_\Omega(X) \to Pr(X)$ satisfies $d(\pi_X \mu, \pi_X \nu) \le d(\mu, \nu)$; the argument uses Lemma 3.22. Thus π_X is Lipschitz with Lipschitz constant

$$[\pi_X]_\mathrm{L} = \sup_{\mu \ne \nu} \frac{d(\pi_X \mu, \pi_X \nu)}{d(\mu, \nu)} = 1,$$

where equality to 1 follows from $\pi_X|_{Pr(X)} = \mathrm{id}$. This does not depend on the choice of the enumeration of the G_m, in particular not on $\Omega = G_0$.

The proof of Theorem 4.16 contains the arguments needed to prove that the narrow topology on $Pr_\Omega(X)$ is generated by the family

$$\{(x, \omega) \mapsto g(x) 1_F(\omega) : g \in \mathrm{BL}(X),\ 0 \le g \le 1,\ [g]_\mathrm{L} \le 1,\ F \in \mathscr{F}\}$$

already, i.e., the narrow topology is the smallest topology such that

$$\mu \mapsto \int_F \int_X g(x)\, d\mu_\omega(x)\, dP(\omega)$$

is continuous for all such $g1_F$. A much shorter and much more direct proof of the metrisability assertion of Theorem 4.16 would be possible by dispersing this argument into a separate proposition. However, then the estimate (4.15) cannot be obtained without spending more effort. We will use (4.15) to deduce completeness of the metric introduced in Theorem 4.16 in the following proposition.

4.18 Proposition *Suppose that* $(\Omega, \mathscr{F}, P)$ *is countably generated* (mod P). *Then the metric* d, *introduced in Theorem 4.16, is complete.*

PROOF Suppose that $(\mu_n)_{n\in\mathbb{N}}$ is a Cauchy sequence in $Pr_\Omega(X)$ with respect to the metric d. Then also $(\pi_X\mu_n)_{n\in\mathbb{N}}$ is a Cauchy sequence in $Pr(X)$ by Remark 4.17 (iii). Furthermore, the metric d reduced to the non-random measures $Pr(X)$ is complete, since it is equivalent to the metric β from (4.32) by Remark 4.17 (ii), which, in turn, is complete (compare, e.g., Corollary 11.5.5 of Dudley [19], p. 317). Thus, the Cauchy sequence $(\pi_X\mu_n)_{n\in\mathbb{N}}$ converges to some $\rho \in Pr(X)$ in the (non-random) narrow topology of $Pr(X)$. Now for $f \in \mathrm{BL}_\Omega(X)$ choose $C \in \mathbb{R}$ with $\|f(\cdot,\omega)\|_{\mathrm{BL}} \leq C$ for P-almost all ω (such a C exists by definition of $\mathrm{BL}_\Omega(X)$, see Definition 4.6). Then $g = (f + C)/2C^2$ satisfies $0 \leq g \leq 1$ and $[g(\cdot,\omega)]_\mathrm{L} \leq 1$ for P-almost all ω. From (4.15) of Theorem 4.16 we get

$$|\mu(g) - \nu(g)| \leq d(\mu,\nu) + 2\sqrt{5}d(\mu,\nu)^{1/2},$$

hence

$$|\mu(f) - \nu(f)| \leq 2C^2\big(d(\mu,\nu) + 2\sqrt{5}d(\mu,\nu)^{1/2}\big)$$

for all $\mu, \nu \in Pr_\Omega(X)$. Consequently, $\big(\mu_n(f)\big)_{n\in\mathbb{N}}$ is a Cauchy sequence of real numbers for every $f \in \mathrm{BL}_\Omega(X)$. Put

$$L(f) = \lim_{n\to\infty} \mu_n(f). \qquad (4.33)$$

Clearly L is linear and nonnegative on $\mathrm{BL}_\Omega(X)$. For any non-random $g \in \mathrm{BL}(X) \subset \mathrm{BL}_\Omega(X)$, Lemma 3.22 yields $\mu_n(g) = \pi_X\mu_n(g)$; hence for non-random g convergence of $\pi_X\mu_n$ to ρ implies $L(g) = \rho(g)$. Furthermore, for any $h \in L^\infty(P)$ we have $\mu_n(h) = Eh$ for all $n \in \mathbb{N}$; hence $L(h) = Eh = \int h \, dP$. Thus L satisfies all assumptions of Proposition 4.12, so there exists a random measure μ with $\mu(g) = L(g)$ for all $g \in \mathrm{BL}_\Omega(X)$. By (4.33), the sequence $(\mu_n)_{n\in\mathbb{N}}$ converges to $\mu \in Pr_\Omega(X)$ in the narrow topology. This holding true for any Cauchy sequence (μ_n) with respect to the metric d, d is complete. □

It might not be clear yet as to whether the assumption of a countably generated σ-algebra is really necessary in order to have the narrow topology on

$Pr_\Omega(X)$ metrisable. It will turn out that it is, except for the trivial case of a one point X, in which case $Pr_\Omega(X)$ is one point itself. Details will be given in Remark 4.30.

Conditional Expectation for Random Measures

We need the notion of the conditional expectation of a random measure with respect to a sub-σ-algebra of $\mathscr{F}$. This will turn out to be useful not only to obtain sequential compactness of tight sets of random measures but also to characterise separability of the narrow topology. For a σ-algebra $\mathscr{E}$ over Ω we will use the notation $\mathscr{E} \subset \mathscr{F}$ (mod P) if for every $E \in \mathscr{E}$ there exists $F \in \mathscr{F}$ with $P(E \bigtriangleup F) = 0$. Equivalently, $\mathscr{E} \subset \mathscr{F}$ (mod P) if $\mathscr{E}$ is contained in the completion of $\mathscr{F}$ with respect to P. In the following, given a bounded measurable or P-integrable $h : \Omega \to \mathbb{R}$ we denote by $E(h \mid \mathscr{E}) : \Omega \to \mathbb{R}$ the usual conditional expectation. It is P-almost everywhere uniquely determined by being measurable with respect to $\mathscr{E}$, and satisfying $\int_E E(h \mid \mathscr{E}) \, dP = \int_E h \, dP$ for every $E \in \mathscr{E}$.

4.19 Definition (CONDITIONAL EXPECTATION FOR RANDOM MEASURES)
Suppose that $(\Omega, \mathscr{F}, P)$ is a probability space, and that $\mathscr{E}$ is a σ-algebra over Ω with $\mathscr{E} \subset \mathscr{F}$.

(i) If μ is a random measure on a Polish space X then the disintegration of the restriction of μ to $\mathscr{B} \otimes \mathscr{E}$ with respect to the restriction of P to $\mathscr{E}$ is said to be the *conditional expectation of μ with respect to $\mathscr{E}$*, and is denoted by $\omega \mapsto E(\mu \mid \mathscr{E})_\omega$. Note that

$$
\begin{aligned}
E(\mu \mid \mathscr{E})_\omega(B) &= E(\mu(B) \mid \mathscr{E})(\omega), \quad \text{and} \\
E(\mu \mid \mathscr{E})_\omega(f) &= E(\mu(f) \mid \mathscr{E})(\omega)
\end{aligned}
\tag{4.34}
$$

for P-almost all ω, for every $B \in \mathscr{B}$ and for every $f : X \to \mathbb{R}$ Borel measurable and bounded. Here $\mu(B)$ and $\mu(f)$ denote the real valued random variables $\omega \mapsto \mu_\omega(B)$ and $\omega \mapsto \mu_\omega(f) = \int f(x) \, d\mu_\omega(x)$, respectively (thus deviating for a moment from the notation introduced in (3.2) after Remark 3.8, which had been used up to now).

(ii) A random measure μ is said to be *measurable with respect to $\mathscr{E}$*, or *$\mathscr{E}$-measurable*, if

$$
\mu_\omega = E(\mu \mid \mathscr{E})_\omega
$$

for P-almost all $\omega \in \Omega$. By (4.34) together with Remark 3.2, this is equivalent to $\mathscr{E}$-measurability of $\omega \mapsto \mu_\omega(B)$ for every $B \in \mathscr{B}$, where we use the fact that the Borel σ-algebra of X is countably generated. Thus $\mathscr{E}$-measurability of $\mu \in Pr_\Omega(X)$ is equivalent to

$$\mu_\omega(B) = E(\mu(B) \,|\, \mathscr{E})(\omega)$$

P-a.s. for every $B \in \mathscr{B}$ (it suffices to have the identity for every B from a countable set of open or closed sets generating the topology of X). This, in turn, is equivalent to

$$\mu_\omega(f) = E(\mu(f) \,|\, \mathscr{E})(\omega) \tag{4.35}$$

P-a.s. for every $f \in C(X)$.

(iii) Denote by

$$Pr_{\Omega,\mathscr{E}}(X) = \big\{ \mu \in Pr_\Omega(X) : \mu \ \mathscr{E}\text{-measurable } P\text{-a.s.}\big\}$$

the set of almost surely $\mathscr{E}$-measurable random measures, and by

$$\pi_\mathscr{E} : Pr_\Omega(X) \to Pr_{\Omega,\mathscr{E}}(X)$$

the conditional expectation $\mu \mapsto E(\mu \,|\, \mathscr{E})$.

For $\mathscr{E} = \{\emptyset, \Omega\}$ the trivial σ-algebra we get $Pr_{\Omega,\mathscr{E}}(X) = Pr(X)$ and $\pi_\mathscr{E}\mu = E(\mu \,|\, \mathscr{E}) = \pi_X \mu$, the projection of μ to its marginal on X.

4.20 Remark For $\mathscr{E} \subset \mathscr{F}$ (mod P) with $\mathscr{F} \neq \mathscr{E}$ (mod P) the set of $\mathscr{E}$-measurable random measures $Pr_{\Omega,\mathscr{E}}(X)$ is a proper subset of $Pr_\Omega(X)$, provided X is not a one point set. In fact, for $x, y \in X$, $x \neq y$, choose $F \in \mathscr{F}$ such that $P(F \triangle E) \neq 0$ for all $E \in \mathscr{E}$, or, equivalently, $1_F \neq E(1_F \,|\, \mathscr{E})$ with positive probability. Then $\omega \mapsto \mu_\omega = 1_F(\omega)\delta_x + (1 - 1_F(\omega))\delta_y$ does not satisfy (4.18) for any $B \in \mathscr{B}$ with $x \in B$ and $y \notin B$. Consequently, $\mu \in Pr_\Omega(X) \setminus Pr_{\Omega,\mathscr{E}}(X)$.

For $h, g \in L^1(\Omega, \mathscr{F}, P)$ with $hg \in L^1(\Omega, \mathscr{F}, P)$ and a σ-algebra $\mathscr{E} \subset \mathscr{F}$ (mod P) it is easily verified that

$$\int h\, E(g \,|\, \mathscr{E})\, dP = \int E(h \,|\, \mathscr{E})\, E(g \,|\, \mathscr{E})\, dP = \int E(h \,|\, \mathscr{E})\, g\, dP. \tag{4.36}$$

4.21 Proposition Let $\mathscr{E} \subset \mathscr{F}$ (mod P) be a σ-algebra. Then $Pr_{\Omega,\mathscr{E}}(X)$ is a closed subset of $Pr_\Omega(X)$ (with respect to the narrow topology of $Pr_\Omega(X)$).

PROOF Suppose that μ^α is a net in $Pr_{\Omega,\mathscr{E}}(X)$ converging to some $\mu \in Pr_\Omega(X)$. We want to prove that $\mu \in Pr_{\Omega,\mathscr{E}}(X)$. This follows from Lemma 3.14 together with (4.35), once we have established

$$\int_F \mu_\omega(g) \, dP(\omega) = \int_F E(\mu(g) \mid \mathscr{E})(\omega) \, dP(\omega) \tag{4.37}$$

for every $g \in C(X)$ and every $F \in \mathscr{F}$.

Convergence of μ^α to μ in the narrow topology of $Pr_\Omega(X)$ implies convergence of the net (of real numbers) $\int_\Omega h\mu^\alpha(g) \, dP$ to $\int_\Omega h\mu(g) \, dP$ for every $h \in L^1(\Omega, \mathscr{F}, P)$ and every $g \in C(X)$. Since μ^α is $\mathscr{E}$-measurable P-a.s., we have

$$\int_\Omega h \, \mu^\alpha(g) \, dP = \int_\Omega h \, E(\mu^\alpha(g) \mid \mathscr{E}) \, dP = \int_\Omega E(h \mid \mathscr{E}) \, \mu^\alpha(g) \, dP,$$

using (4.35) for the first and (4.36) for the second identity. Narrow convergence of μ^α to μ, evaluated for the random continuous function $(x, \omega) \mapsto g(x)E(h \mid \mathscr{E})(\omega)$, implies that $\int_\Omega E(h \mid \mathscr{E}) \, \mu^\alpha(g) \, dP$ converges to

$$\int_\Omega E(h \mid \mathscr{E}) \, \mu(g) \, dP = \int_\Omega h \, E(\mu(g) \mid \mathscr{E}) \, dP,$$

using (4.36) again. Consequently, the integrals $\int_\Omega h \, \mu^\alpha(g) \, dP$ converge with α to $\int_\Omega h \, \mu(g) \, dP$, and they also converge with α to $\int_\Omega h \, E(\mu(g) \mid \mathscr{E}) \, dP$, whence

$$\int_\Omega h \, \mu(g) \, dP = \int_\Omega h \, E(\mu(g) \mid \mathscr{E}) \, dP.$$

This holds true for every $h \in L^1(\Omega, \mathscr{F}, P)$ and every $g \in C(X)$. With $h = 1_F$ for $F \in \mathscr{F}$ we get (4.37), which implies $\mu = E(\mu \mid \mathscr{E})$ P-a.s. $\qquad\square$

The next Corollary is immediate by taking $\mathscr{E} = \{\emptyset, \Omega\}$.

4.22 Corollary *The set $Pr(X)$ of non-random probability measures on X is a closed subset of $Pr_\Omega(X)$ with respect to the narrow topology of $Pr_\Omega(X)$.*

To obtain another consequence of Proposition 4.21 we use an elementary fact. The argument needs the set of (non-random) probability measures $Pr(X)$ to be separable. For X Polish this is satisfied, since $Pr(X)$ is Polish itself. In order to keep the structure transparent we formulate the argument as a lemma.

4.23 Lemma *Suppose that D is a countable subset of $Pr_\Omega(X)$. Then there exists a countably generated (mod P) σ-algebra $\mathscr{E}$ with $D \subset Pr_{\Omega,\mathscr{E}}(X)$.*

PROOF Any random measure $\mu \in Pr_\Omega(X)$ is a random variable taking values in the space $Pr(X)$ of probability measures on X equipped with the Borel σ-algebra $\mathscr{B}(Pr(X))$ of the narrow topology (cf. Remark 3.20 (i)). Since $Pr(X)$ with the narrow topology is Polish, its Borel σ-algebra $\mathscr{B}(Pr(X))$ is countably generated. Consequently, the σ-algebra $\mu^{-1}(\mathscr{B}(Pr(X)))$ generated by an arbitrary $\mu \in Pr_\Omega(X)$ on Ω is countably generated. For the countable $D \subset Pr_\Omega(X)$ put $\mathscr{E} = \sigma\{\mu : \mu \in D\} \subset \mathscr{F}$. Then $\mathscr{E}$ is countably generated, and clearly $\mu \in Pr_{\Omega,\mathscr{E}}(X)$ for every $\mu \in D$.   □

4.24 Corollary *Suppose that X is not a one point set. Then the narrow topology on $Pr_\Omega(X)$ cannot be separable unless $\mathscr{F}$ is countably generated (mod P).*

PROOF Suppose that $D \subset Pr_\Omega(X)$ is countable and dense with respect to the narrow topology. Let $\mathscr{E} = \sigma\{\mu \in D\}$ be the initial σ-algebra of D, which is countably generated. Then $D \subset Pr_{\Omega,\mathscr{E}}(X)$. By Proposition 4.21, therefore also the closure of D is contained in $Pr_{\Omega,\mathscr{E}}(X)$, which implies $Pr_\Omega(X) = Pr_{\Omega,\mathscr{E}}(X)$. By Remark 4.20, this can hold only if $\mathscr{E} = \mathscr{F}$ (mod P). Therefore $\mathscr{F}$ must have been countably generated (mod P).   □

We now take up the considerations of Remarks 3.21 and 3.23 for the more general $Pr_{\Omega,\mathscr{E}}(X)$ instead of $Pr(X)$. It should be emphasised that the situation is slightly different now, insofar as we have introduced $Pr_{\Omega,\mathscr{E}}(X)$ as a subspace of $Pr_\Omega(X)$, whereas $Pr(X)$ had come up as a space in its own right, which we then interpreted as a subspace of $Pr_\Omega(X)$.

4.25 Remark For a sub-σ-algebra $\mathscr{E}$ of $\mathscr{F}$ (mod P), Definition 4.19 introduced the set $Pr_{\Omega,\mathscr{E}}(X)$ of $\mathscr{E}$-measurable random measures as a subset of $Pr_\Omega(X)$. Then $\pi_\mathscr{E} : Pr_\Omega(X) \to Pr_{\Omega,\mathscr{E}}(X) \subset Pr_\Omega(X)$, $\mu \mapsto \pi_\mathscr{E}\mu = E(\mu \,|\, \mathscr{E})$, satisfies $\pi_\mathscr{E}^2 = \pi_\mathscr{E}$. Clearly $\pi_\mathscr{E}(Pr_\Omega(X)) = Pr_{\Omega,\mathscr{E}}(X)$, so Proposition B.1 applies, yielding that the quotient topology on $Pr_{\Omega,\mathscr{E}}(X)$ induced by $\pi_\mathscr{E}$ is weaker than the trace of the narrow topology on $Pr_{\Omega,\mathscr{E}}(X)$.

We can consider $Pr_{\Omega,\mathscr{E}}(X)$ also as a space in its own right, with its own narrow topology generated by $C_{\Omega,\mathscr{E}}(X)$, the set of all random continuous functions such that $\omega \mapsto f(x,\omega)$ is measurable with respect to $\mathscr{E}$ for every $x \in X$. In order to investigate the relations between the different topologies on $Pr_{\Omega,\mathscr{E}}(X)$ we need the following observation.

4.26 Proposition *For any σ-algebra $\mathscr{E} \subset \mathscr{F}$ (mod P),*

$$\int_\Omega \int_X f(x,\omega)\,d(\pi_\mathscr{E}\mu)_\omega(x)\,dP(\omega)$$

$$= \int_\Omega \int_X E(f(x,\cdot)\,|\,\mathscr{E})(\omega)\,d(\pi_\mathscr{E}\mu)_\omega(x)\,dP(\omega)$$

$$= \int_\Omega \int_X E(f(x,\cdot)\,|\,\mathscr{E})(\omega)\,d\mu_\omega(x)\,dP(\omega) \tag{4.38}$$

for every $\mu \in Pr_\Omega(X)$, and for every $f : X \times \Omega \to \mathbb{R}$ bounded and $\mathscr{B} \otimes \mathscr{F}$-measurable.

PROOF For $g : X \to \mathbb{R}$ bounded and measurable, (4.35) yields $\pi_\mathscr{E}\mu(g) = E(\mu\,|\,\mathscr{E})(g) = E(\mu(g)\,|\,\mathscr{E})$, using the notation $\mu(g)$ for the random variable $\omega \mapsto \mu_\omega(g)$. For $h : \Omega \to \mathbb{R}$ bounded and $\mathscr{F}$-measurable we thus get, invoking (4.36),

$$\int h\,E(\mu(g)\,|\,\mathscr{E})\,dP = \int E(h\,|\,\mathscr{E})E(\mu(g)\,|\,\mathscr{E})\,dP = \int E(h\,|\,\mathscr{E})\,\mu(g)\,dP.$$

This gives the assertion for $f(x,\omega) = g(x)h(\omega)$. Put

$$\mathcal{H} = \{f : X \times \Omega \to \mathbb{R} : f \text{ bounded measurable, and satisfies (4.38)}\}.$$

Then $\mathcal{H}$ is a vector space, and for every $A = B \times F \in \mathscr{B} \times \mathscr{F}$ we have $(x,\omega) \mapsto 1_A(x,\omega) = 1_B(x)\,1_F(\omega) \in \mathcal{H}$. For $0 \le f_n \in \mathcal{H}$, $n \in \mathbb{N}$, with $f_n \nearrow f$ for $f : X \times \Omega \to \mathbb{R}$ bounded then $E(f_n\,|\,\mathscr{E}) \nearrow E(f\,|\,\mathscr{E})$ for ν-almost all (x,ω) for every $\nu \in Pr_\Omega(X)$, so $f \in \mathcal{H}$ by monotone convergence. Consequently, $\mathcal{H}$ contains all bounded functions measurable with respect to $\sigma(\mathscr{B} \times \mathscr{F}) = \mathscr{B} \otimes \mathscr{F}$ by a monotone class theorem (Williams [39], Theorem II.4, p. 40). □

Note that (4.38) is a generalisation of (3.10). The main additional difficulty in establishing (4.38) comes from the fact that the first identity of (3.10), which followed from Fubini, has no correspondence here. Writing $E(\mu\,|\,\mathscr{E})$ instead of $\pi_\mathscr{E}\mu$, (4.38) is on the other hand very similar to (4.36), giving

$$\int_\Omega \int_X f(x,\omega)\,dE(\mu\,|\,\mathscr{E})_\omega(x)\,dP(\omega)$$

$$= \int_\Omega \int_X E(f(x,\cdot)\,|\,\mathscr{E})(\omega)\,d\mu_\omega(x)\,dP(\omega)$$

$$= \int_\Omega \int_X E(f(x,\cdot)\,|\,\mathscr{E})(\omega)\,dE(\mu\,|\,\mathscr{E})_\omega(x)\,dP(\omega).$$

4.27 Corollary *The map* $\pi_\mathscr{E} : Pr_\Omega(X) \to Pr_\Omega(X)$ *is continuous with respect to the narrow topology on* $Pr_\Omega(X)$. *In particular, the following three topologies on* $Pr_{\Omega,\mathscr{E}}(X)$ *coincide:*

(i) *the narrow topology on* $Pr_{\Omega,\mathscr{E}}(X)$ *as a space of random measures in its own right,*

(ii) *the trace of the narrow topology of* $Pr_\Omega(X)$ *on* $Pr_{\Omega,\mathscr{E}}(X) \subset Pr_\Omega(X)$,

(iii) *the quotient topology induced by* $\pi_\mathscr{E} : Pr_\Omega(X) \to Pr_{\Omega,\mathscr{E}}(X)$.

PROOF By Corollary 4.10 it suffices to prove that

$$\mu \mapsto \pi_\mathscr{E}\mu(g)$$

is continuous for every $g \in \mathrm{BL}_\Omega(X)$. From Proposition 4.26 we get

$$\pi_\mathscr{E}\mu(g) = \int_\Omega \int_X E\big(g(x,\cdot) \,|\, \mathscr{E}\big)\, d\mu_\omega(x)\, dP(\omega). \qquad (4.39)$$

For $x \in X$ fixed put $b(x,\omega) = E\big(g(x,\cdot)\,|\,\mathscr{E}\big)(\omega)$. Let $D \subset X$ be countable and dense. Then, for all ω outside a P-nullset depending only on D, the function $x \mapsto b(x,\omega)$ is a bounded Lipschitz function on D. By Theorem 6.1.1 of Dudley ([19], p. 146), for every of these ω's $b(\cdot,\omega)$ can be extended to a bounded Lipschitz function on all of X. The extension is unique, since D is dense in X, so $b \in \mathrm{BL}_\Omega(X)$. Furthermore, for every $x \in X$ the function $\omega \mapsto b(x,\omega)$ satisfies $b(x,\cdot) = E\big(g(x,\cdot)\,|\,\mathscr{E}\big)$ for P-almost all ω (where the nullset may depend on x). Therefore, (4.39) yields $\pi_\mathscr{E}\mu(g) = \mu\big(E(g\,|\,\mathscr{E})\big) = \mu(b)$. From $b \in \mathrm{BL}_\Omega(X)$ we get that $\mu \mapsto \pi_\mathscr{E}(g) = \mu(b)$ is continuous.

In order to establish identity of the three topologies, first note that the narrow topology on $Pr_{\Omega,\mathscr{E}}(X)$ is clearly weaker than the trace of the narrow topology on $Pr_\Omega(X)$. Continuity of $\pi_\mathscr{E}$, together with the fact that the restriction of $\pi_\mathscr{E}$ to $Pr_{\Omega,\mathscr{E}}(X)$ is the identity, implies that the two topologies coincide. Finally, using continuity of $\pi_\mathscr{E}$ again, we get coincidence of the trace of the narrow topology of $Pr_\Omega(X)$ on $Pr_{\Omega,\mathscr{E}}(X)$ with the quotient topology induced by $\pi_\mathscr{E}$ on $Pr_{\Omega,\mathscr{E}}(X)$ from Corollary B.2. Note that here continuity of $\pi_\mathscr{E}$ with respect to the trace topology suffices, which is immediate from the second identity of (4.38) already. $\square$

The Prohorov Theorem for the Narrow Topology on Random Measures

We are now in a position to prove that compactness implies sequential compactness with respect to the narrow topology on the space of random measures, even if the σ-algebra of the probability space is not countably generated.

4.28 Proposition *If $\Gamma \subset Pr_\Omega(X)$ is compact then it is also sequentially compact, both with respect to the narrow topology of $Pr_\Omega(X)$.*

PROOF For $(\mu_n)_{n \in \mathbb{N}} \subset \Gamma$ put $\mathscr{E} = \sigma\{\mu_n : n \in \mathbb{N}\}$. Then $\mathscr{E}$ is countably generated, so the narrow topology on $Pr_{\Omega, \mathscr{E}}(X)$ is metrisable by Theorem 4.16. Since $\pi_\mathscr{E}$ is continuous by Corollary 4.27, with Γ also $\pi_\mathscr{E}(\Gamma)$ is compact in the metrisable narrow topology of $Pr_{\Omega, \mathscr{E}}$. Now $\pi_\mathscr{E} \mu_n = \mu_n \in Pr_{\Omega, \mathscr{E}}(X)$ for all $n \in \mathbb{N}$, so the sequence μ_n has a subsequence which converges with respect to the metrisable narrow topology of $Pr_{\Omega, \mathscr{E}}(X)$. But this topology coincides, again by Corollary 4.27, with the trace of the narrow topology of $Pr_\Omega(X)$. Therefore the subsequence converges also with respect to the narrow topology of $Pr_\Omega(X)$. $\square$

This completes the proof of Theorem 4.4. We state it again, collecting the arguments constituting its proof.

4.29 Theorem (PROHOROV THEOREM FOR RANDOM MEASURES; IDENTICAL WITH THEOREM 4.4) *Suppose that $\Gamma \subset Pr_\Omega(X)$. Then Γ is tight if, and only if, it is relatively compact with respect to the narrow topology of $Pr_\Omega(X)$. Furthermore, in this case Γ is relatively sequentially compact, again with respect to the narrow topology of $Pr_\Omega(X)$.*

PROOF Relative compactness of Γ implies tightness by Lemma 4.5. Tightness of Γ implies relative compactness of Γ by Corollary 4.15. Relative compactness of Γ implies sequential relative compactness by Proposition 4.28. $\square$

4.30 Remark If in the situation of Theorem 4.16 the σ-algebra $\mathscr{F}$ is not countably generated (mod P), then the narrow topology on $Pr_\Omega(X)$ is not metrisable (provided X is not a one point set; in this case $Pr_\Omega(X)$ is a one point set as well). In fact, pick $x, y \in X$, $x \neq y$. Then $\{x, y\} \subset X$ is closed, and $E = \{\mu \in Pr_\Omega(X) : \mu_\omega\{x, y\} = 1 \ P\text{-a.s.}\}$ is a closed subset of $Pr_\Omega(X)$ by Corollary 3.18. Denote by $L^\infty(\Omega, \mathscr{F}, P) = L^\infty(\Omega, P)$ the usual space of

P-essentially bounded real valued $h : \Omega \to \mathbb{R}$, identifying P-a.s. equal ones. Then $\xi : E \to L^\infty(\Omega, P)$, $\xi(\mu)(\omega) = \mu_\omega\{x\}$, defines an injective map from E onto $\{f \in L^\infty(\Omega, P) : 0 \leq f \leq 1\}$. Furthermore, ξ is a homeomorphism with respect to the (trace of the) narrow topology on E and the (trace of the) weak* topology (obtained by interpreting L^∞ as the dual Banach space of $L^1(\Omega, \mathscr{F}, P)$). Now $\{f \in L^\infty(\Omega, P) : 0 \leq f \leq 1\}$ is homeomorphic to the unit ball of $L^\infty(\Omega, P)$, both in the strong L^∞ and in the weak* topology (with a homeomorphism given by $f \mapsto 2f - 1$). By Theorem V.5.1 of Dunford and Schwartz ([20], p. 426), the unit ball of the dual of a Banach space (here: $L^1(\Omega, \mathscr{F}, P)$) is metrisable if and only if the Banach space is separable. Since $L^1(\Omega, \mathscr{F}, P)$ is separable if and only if $\mathscr{F}$ is countably generated (mod P) (see, e.g., Halmos [25], Exercise (1), p. 177, and § 40 Theorem B, p. 168), the trace of the narrow topology on E cannot be metrisable unless $\mathscr{F}$ is countably generated (mod P). But if the trace of the narrow topology on $E \subset Pr_\Omega(X)$ is not metrisable, then the narrow topology cannot be metrisable on all of $Pr_\Omega(X)$ either. We summarise Theorem 4.16 together with this remark in the next corollary.

4.31 Corollary *The narrow topology on the space $Pr_\Omega(X)$ of random probability measures on a Polish space X with at least two points over a probability space $(\Omega, \mathscr{F}, P)$ is metrisable if and only if $\mathscr{F}$ is countably generated (mod P).*

4.32 Remark We had been embedding X into $Pr(X)$ via $x \mapsto \delta_x$, and $Pr(X)$ into $Pr_\Omega(X)$ by interpreting a non-random measure ρ as a constant $Pr(X)$-valued random variable. There is yet another embedding, making the square full. Denote by $X_\Omega = \{x : \Omega \to X : x \ (\mathscr{F}, B)\text{-measurable}\}$ the space of measurable maps from Ω to X. Then X can be considered a subset of X_Ω by interpreting $x \in X$ as the constant $\omega \mapsto x$. In turn, X_Ω can be embedded into $Pr_\Omega(X)$ by mapping $\omega \mapsto x(\omega)$ to $\omega \mapsto \delta_{x(\omega)}$. We thus obtain the following diagram, with all maps the corresponding embeddings, trivially commuting.

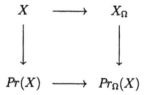

The arrow from $Pr(X)$ to $Pr_\Omega(X)$ can be inverted by π_X. However, X is not a linear space in general, so in general we neither get an arrow from X_Ω to X nor one from $Pr(X)$ to X. Random measures of the form $\omega \mapsto \delta_{x(\omega)}$ are addressed to as *Young measures* by Valadier [37]. The trace of the narrow topology of $Pr_\Omega(X)$ on the Young measures provides a topology on X_Ω.

The previous remark can be extended to a big, though not very substantial, commuting diagram. Let $\mathscr{E}_1$ and $\mathscr{E}_2$ be two σ-algebras on Ω with $\mathscr{E}_1 \subset \mathscr{E}_2 \subset \mathscr{F}$ (mod P), and denote by $X_{\Omega,\mathscr{E}_1}$ and $X_{\Omega,\mathscr{E}_2}$, respectively, the spaces of maps $x : \Omega \to X$ which are $\mathscr{E}_1$- and $\mathscr{E}_2$-measurable, respectively. Then

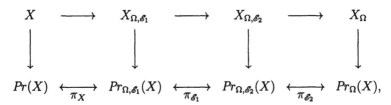

where all one directional arrows and all arrows directed to the right are embeddings, and the arrows going to the left are given by the projections π_*, $* \in \{X, \mathscr{E}_1, \mathscr{E}_2\}$. Clearly $\pi_{\mathscr{E}_1} \circ \pi_{\mathscr{E}_2} = \pi_{\mathscr{E}_1}$, and $\pi_X \circ \pi_{\mathscr{E}_1} = \pi_X$.

Chapter 5

Further Topologies on Random Measures

In the present chapter we discuss further possibilities to topologise the space $Pr_\Omega(X)$ and their relations with the narrow topology. Let $(\Omega, \mathscr{F}, P)$ be a probability space, and let X be a Polish space.

Topologies on Random Variables in Metric Spaces

We will here consider random measures $\mu \in Pr_\Omega(X)$ as random variables $\mu : \Omega \to Pr(X)$ with values in the Polish space $Pr(X)$, equipped with the Borel σ-algebra of the narrow topology, (see Remark 3.20 (i)). Choose a metric d on the Polish space $Pr(X)$ of non-random probability measures. Then $\omega \mapsto d(\mu_\omega, \nu_\omega)$ is a measurable nonnegative real valued function by Remark 3.20 (ii). Put

$$\beta(\mu, \nu) = \int_\Omega d(\mu_\omega, \nu_\omega) \, dP(\omega).$$

Clearly β is a metric on $Pr_\Omega(X)$ (recall that $\omega \mapsto \mu_\omega$ and $\omega \mapsto \nu_\omega$ are identified if they coincide P-a.s.).

More generally, for $1 \le p \le \infty$ put

$$\beta_p(\mu, \nu) = \begin{cases} \left(\int_\Omega d(\mu_\omega, \nu_\omega)^p \, dP(\omega) \right)^{1/p} = \|d(\mu_\omega, \nu_\omega)\|_p & \text{for } 1 \le p < \infty, \\ \|d(\mu_\omega, \nu_\omega)\|_\infty & \text{for } p = \infty, \end{cases}$$

67

where $\| \cdot \|_\infty$ denotes the L^∞-norm on $L^\infty(\Omega, \mathscr{F}, P)$. Then also β_p defines a metric for every p, $1 \le p \le \infty$.

Another metric is given by the Ky Fan metric on the space of $Pr(X)$-valued random variables,

$$\varkappa(\mu, \nu) = \inf\{\varepsilon > 0 : P\{\omega : d(\mu_\omega, \nu_\omega) > \varepsilon\} \le \varepsilon\}$$

for $\mu, \nu \in Pr_\Omega(X)$. The Ky Fan metrises convergence in probability in the sense that a sequence $\mu^n \in Pr_\Omega(X)$, $n \in \mathbb{N}$, converges to $\mu \in Pr_\Omega(X)$ in probability (as random variables on the metric space $Pr(X)$) if and only if $\varkappa(\mu^n, \mu)$ converges to 0 with $n \to \infty$ (Dudley [19], Theorem 9.2.2, p. 227).

For any p_1, p_2 with $1 \le p_1 \le p_2 \le \infty$ we have

$$\varkappa \le \beta_1^{1/2} \quad \text{and} \quad \beta_1 \le \beta_{p_1} \le \beta_{p_2} \le \beta_\infty, \tag{5.1}$$

see Lemma B.5 for the first inequality. Of course, β_p and $\varkappa$ depend on the choice of the metric d on $Pr(X)$, so (5.1) in general only holds for β_p and $\varkappa$ defined with respect to the same d. For each choice of d on $Pr(X)$ we obtain a family of topologies on $Pr_\Omega(X)$ which is increasing by (5.1), the strongest being the topology induced by β_∞ and the weakest the one induced by the Ky Fan metric $\varkappa$. As soon as the metric d on $Pr(X)$ is complete, also the Ky Fan metric $\varkappa$ is complete (Dudley [19], Theorem 9.2.3, p. 227). Since every Cauchy sequence in β_p, $1 \le p \le \infty$, is a Cauchy sequence in the Ky Fan metric by (5.1), also all the β_p are complete then.

Next, one may ask when a sequence $\mu^n \in Pr_\Omega(X)$, $n \in \mathbb{N}$, converges to $\mu \in Pr_\Omega(X)$ P-a.s. It is known that almost sure convergence can be metrised if and only if it coincides with convergence in probability (see, for example, Dudley [19], Remark after Theorem 9.2.2, p. 227), and that in general there is no topology on the set of measurable maps from a probability space to a separable metric space such that almost sure convergence is convergence in this topology (Dudley [19] § 9.2, Problem 2, p. 228).

Finally, one can consider convergence in law on $Pr_\Omega(X)$. The distribution or law of $\mu \in Pr_\Omega(X)$ is the image measure of P under the map $\mu : \Omega \to Pr(X)$, denoted by μP or $\mathcal{L}(\mu)$. It is a probability measure on $Pr(X)$ equipped with the Borel σ-algebra associated with the (non-random) narrow topology on $Pr(X)$ (compare Remark 3.20 (i)), so $\mathcal{L}(\mu) \in Pr(Pr(X))$. A sequence $(\mu^n)_{n \in \mathbb{N}} \in Pr_\Omega(X)$ converges to $\mu \in Pr_\Omega(X)$ in law (or: in distribution) if the laws $\mathcal{L}(\mu^n) = \mu^n P$ converge to the law $\mathcal{L}(\mu) = \mu P$ in the narrow topology of $Pr(Pr(X))$. The notion of convergence in distribution/law does not refer to a choice of a metric on $Pr(X)$. The laws of distinct random measures can coincide; so convergence in law cannot be induced by a Hausdorff topology on $Pr_\Omega(X)$. If we equip the space of laws on $Pr(X)$ with the Prohorov metric

$$\alpha(\Lambda, \Sigma) = \inf\{r > 0 : \Lambda(B) \leq \Sigma(B^r) + r \text{ for all } B \in \mathscr{B}(Pr(X))\}$$

for $\Lambda, \Sigma \in Pr(Pr(X))$, then for any two random measures μ and ν

$$\alpha(\mathcal{L}(\mu), \mathcal{L}(\nu)) \leq \varkappa(\mu, \nu) \tag{5.2}$$

(see Dudley [19], Theorem 11.3.5, p. 312). The Prohorov metric α topologises the narrow topology on $Pr(Pr(X))$ (Theorem A.2), so the map $\mathcal{L} : Pr_\Omega(X) \to Pr(Pr(X))$ is continuous with respect to the topology induced by $\varkappa$ on $Pr_\Omega(X)$ and the narrow topology on $Pr(Pr(X))$. In particular, convergence of random measures in probability implies their convergence in law. Of course, for sequences this is well-known also without invoking (5.2) (see Dudley [19], Theorem 9.3.5, p. 231, or Gänssler and Stute [24], Satz 8.2.4, p. 333–334). In view of the fact that the distribution of two distinct random measures can coincide, convergence in law cannot imply convergence in probability. However, as soon as a sequence $(\mu^n)_{n\in\mathbb{N}}$ in $Pr_\Omega(X)$ converges in law to a non-random measure $\rho \in Pr(X)$, which means that $\mathcal{L}(\mu^n)$ converges to the Dirac measure δ_ρ in $Pr(Pr(X))$, then μ^n converges already in probability to ρ, see Gänssler and Stute [24] Satz 1.12.4, p. 65 (where the assertion is derived for $\mathbb{R}$-valued random variables, using arguments valid in a metric space instead of $\mathbb{R}$).

5.1 Remark The topologies induced by β_p, $1 \leq p \leq \infty$, on $Pr_\Omega(X)$ do depend on the choice of the metric d on $Pr(X)$. In fact, if d is an unbounded metric metrising the narrow topology on $Pr(X)$, then putting $d_0 = \min\{1, d\}$ gives a bounded metric metrising the same topology (Dudley [19], Proposition 2.4.3, p. 38). Now choose a fixed $\rho \in Pr(X)$, a sequence of measurable sets $F_n \in \mathscr{F}$ with $P(F_n)$ tending to zero, and a sequence $\zeta_n \in Pr(X)$ with $d(\rho, \zeta_n) \geq P(F_n)^{-1}$, and let μ^n be the sequence of random measures given by $\mu^n_\omega = 1_{F^c_n}(\omega)\rho + 1_{F_n}(\omega)\zeta_n$. Then $\int_\Omega d_0(\mu^n, \rho)\, dP = P(F_n)$ converges to zero with $n \to \infty$, whereas $\int_\Omega d(\mu^n, \rho)\, dP \geq 1$ for all $n \in \mathbb{N}$. Thus μ^n converges to ρ in the β-metric associated with d_0, whereas it does not converge in the β-metric associated with d.

Clearly, if d_1 and d_2 are metrics on $Pr(X)$ such that $d_1 \leq Cd_2^q$ for some $q \geq 1$, then the topology on $Pr_\Omega(X)$ induced by the β-metric associated with d_2 is stronger than the β-metric associated with d_1. In particular, two metrics on $Pr(X)$, which are equivalent in the sense that either of them is bounded by a multiple of the other, give the same β-topologies on $Pr_\Omega(X)$, $1 \leq p \leq \infty$. It is not hard to see that also the topologies induced by the respective Ky Fan metrics associated with two equivalent metrics coincide. But here a stronger statement holds, see Lemma 5.3.

5.2 Lemma *Suppose that d is a bounded metric on $Pr(X)$. Then the topology induced by any of the metrics β_p, $1 \leq p < \infty$, coincides with the topology induced by the Ky Fan metric $\varkappa$ on $Pr_\Omega(X)$.*

PROOF The topology induced by the Ky Fan metric $\varkappa$ is weaker than each of the topologies induced by β_p for $1 \leq p \leq \infty$ by (5.1). On the other hand, since all topologies under consideration are metrisable, it suffices to show that any convergent sequence with respect to the Ky Fan metric also converges with respect to β_p, $1 \leq p < \infty$. So suppose that $(\mu^n)_{n \in \mathbb{N}}$ is a sequence in $Pr_\Omega(X)$ which converges with respect to the Ky Fan metric to $\mu \in Pr_\Omega(X)$. This means that the sequence of real valued random variables $(\omega \mapsto d(\mu_\omega^n, \mu_\omega))_{n \in \mathbb{N}}$ converges to 0 in probability. Since d is bounded, the set of random variables $\{\omega \mapsto d(\mu_\omega^n, \mu_\omega) : n \in \mathbb{N}\}$ is bounded, hence uniformly integrable; so the sequence $(d(\mu^n, \mu))_{n \in \mathbb{N}}$ converges to 0 in L^p for every $1 \leq p < \infty$ (see Gänssler and Stute [24], Satz 1.14.9, pp. 75–76). But this means that $\beta_p(\mu^n, \mu)$ converges to 0, so μ^n converges to μ also with respect to each of the β_p, $1 \leq p < \infty$.   □

It is clear that the topology induced by β_∞ is strictly stronger than any β_p-topology, $p < \infty$, regardless of the choice of d, provided $(\Omega, \mathscr{F}, P)$ is not finite.

5.3 Lemma *For any sequence $(\mu^n)_{n \in \mathbb{N}}$ and μ in $Pr_\Omega(X)$ the following are equivalent.*

  (i) $\displaystyle \lim_{n \to \infty} \varkappa(\mu^n, \mu) = 0$,

  (ii) $\displaystyle \lim_{n \to \infty} \int_\Omega |F(\mu_\omega^n) - F(\mu_\omega)| \, dP(\omega) = 0$

  for every $F \in C(Pr(X))$, where $C(Pr(X))$ denotes the space of bounded continuous real-valued functions on $Pr(X)$,

regardless of the choice of the metric d on $Pr(X)$. In particular, the topology induced by the Ky Fan metric $\varkappa$ on $Pr_\Omega(X)$ does not depend on the choice of the metric d on $Pr(X)$.

PROOF Equivalence of (i) and (ii) is one of the assertions of Satz 8.2.4 of Gänssler and Stute [24], pp. 333–334. Independence of the topology induced by the respective Ky Fan metrics on the choice of metric d follows since both are metric topologies, so they coincide if they have the same convergent sequences.   □

Topologies on Random Measures

Up to now all considerations did not use the particular fact that we are deal-ing with random measures. What we had been discussing was concerned with random variables taking values in a metric space, which here happened to be $Pr(X)$. The additional structure of $Pr(X)$ as the space of Borel probability measures on X comes into play as soon as we are looking at the different notions of convergence for sequences

$$\left(\omega \mapsto \int_X f(x)\, d\mu_\omega^n(x)\right)_{n \in \mathbb{N}}$$

for $\mu^n \in Pr_\Omega(X)$ and $f \in C(X)$. We will return to this point later, and will now investigate the relations between the above topologies on $Pr_\Omega(X)$ with the narrow topology on $Pr_\Omega(X)$.

5.4 Proposition *The topology induced by the Ky Fan metric is stronger than the narrow topology on $Pr_\Omega(X)$.*

PROOF Since the topology induced by the Ky Fan metric is independent of the choice of the metric d on $Pr(X)$ by Lemma 5.3, we may suppose that $\varkappa$ is defined with respect to the metric $d(\rho, \zeta) = \sup\{\rho(g) - \zeta(g) : g \in \mathrm{BL}(X),\ 0 \le g \le 1, [g]_\mathrm{L} \le 1\}$ on $Pr(X)$ (which is, up to a constant, the same metric as (4.31) discussed in Remark 4.17 (ii) – but note that here we do not assume $\mathscr{F}$ to be countably generated (mod P), so d need not be the restriction of a metric on $Pr_\Omega(X)$). Then for $\mu \in Pr_\Omega(X)$, $\varepsilon > 0$, and $f \in \mathrm{BL}_\Omega(X)$ with $0 \le f \le 1$ and $[f(\cdot, \omega)]_\mathrm{L} \le 1$ for P-almost all ω, we have

$$\left| \int_X f(x, \omega)\, d\mu_\omega(x) - \int_X f(x, \omega)\, d\nu_\omega(x) \right| \le d(\mu_\omega, \nu_\omega) \quad \text{for } P\text{-almost every } \omega,$$

for every $\nu \in Pr_\Omega(X)$. Thus, for every $\nu \in Pr_\Omega(X)$ with $\varkappa(\mu, \nu) < \varepsilon/2$,

$$|\mu(f) - \nu(f)| \le P\left\{\omega : d(\mu_\omega, \nu_\omega) > \frac{\varepsilon}{2}\right\} + \int_{\{\omega : d(\mu_\omega, \nu_\omega) \le \varepsilon/2\}} d(\mu_\omega, \nu_\omega)\, dP(\omega)$$

$$< \frac{\varepsilon}{2} + \frac{\varepsilon}{2} = \varepsilon$$

(using $\rho(g) - \zeta(g) \le 1$ for any $\rho, \zeta \in Pr(X)$ and g measurable with $0 \le g \le 1$). Since the narrow topology on $Pr_\Omega(X)$ is generated by $\mu \mapsto \mu(f)$

with f as above by Corollary 4.10, every open neighbourhood of an arbitrary $\mu \in Pr_\Omega(X)$ contains a $\varkappa$-ball, whence the topology induced by $\varkappa$ is stronger than the narrow topology on $Pr_\Omega(X)$.   □

We will see in Example 5.7 that the topology induced by the Ky Fan metric is strictly stronger than the narrow topology on $Pr_\Omega(X)$.

Concerning separability of the distinct topologies on $Pr_\Omega(X)$, and, in view of Proposition 5.4, in particular concerning separability of the narrow topology, we have the following lemma.

5.5 Lemma *Suppose that the σ-algebra $\mathscr{F}$ of the probability space $(\Omega, \mathscr{F}, P)$ is countably generated (mod P). Then the topology induced by any of the metrics β_p, $1 \le p < \infty$, on $Pr_\Omega(X)$ is separable. Furthermore, the topology induced by the Ky Fan metric $\varkappa$ as well as the narrow topology is separable.*

PROOF   Choose an increasing sequence of partitions $\{F_k^n : k \le n\}$ with $\sigma\{F_k^n : k = 1, \ldots, n,\ n \in \mathbb{N}\} = \mathscr{F}$ (mod P). Since $Pr(X)$ is Polish we can choose a countable dense set $D \subset Pr(X)$. Then

$$\left\{ \sum_{k=1}^{n} 1_{F_k^n}(\omega)\rho_n : \rho_n \in D,\ n \in \mathbb{N} \right\}$$

is dense in any of the topologies induced by β_p, $1 \le p < \infty$ (where again the sum is to be understood as an abbreviating notation). This is proved by essentially the same arguments used to show that the space $L^p(\Omega, \mathscr{F}, P; E)$ of p-integrable random variables taking values in a separable Banach space E is separable, provided the σ-algebra is countably generated (Behrends [7], Satz IV.1.5, p. 160; Dunford and Schwartz [20], Exercise III.9.6, p. 169). Finally, a dense set in a β_p-topology is also dense in the $\varkappa$- and in the narrow topology, since both are weaker than any of the β_p-topologies.   □

In summarising we arrive at a characterisation of the narrow topology on random measures. Here X is said to be *trivial* if it consists of just one point.

5.6 Theorem *Suppose that X is a nontrivial Polish space. Then the space $Pr_\Omega(X)$ of random measures equipped with the narrow topology is itself a Polish space if and only if the σ-algebra $\mathscr{F}$ of the underlying probability space $(\Omega, \mathscr{F}, P)$ is countably generated (mod P). Furthermore, as soon as $\mathscr{F}$ is not countably generated (mod P), the narrow topology on $Pr_\Omega(X)$ is neither metrisable nor separable.*

PROOF If $\mathscr{F}$ is countably generated (mod P) then the narrow topology on $Pr_\Omega(X)$ is metrisable by a complete metric by Theorem 4.16 together with Proposition 4.18, and it is separable by Lemma 5.5. If $\mathscr{F}$ is not countably generated (mod P) then the narrow topology is not metrisable by Corollary 4.31, and it is not separable by Corollary 4.24. □

Comparison of Different Topologies

In order to illustrate the differences between the topologies we take the simplest nontrivial Polish space X, which is the two point set $X = \{a, b\}$, $a \neq b$, with the discrete topology. (It makes the presentation clearer not to choose $a = 0$ and $b = 1$.) All function spaces (measurable, continuous, Lipschitz) coincide and are equal to $\mathbb{R}^X$, which can be identified with $\mathbb{R}^2$. We will denote functions $f : \{a, b\} \times \Omega \to \mathbb{R}$ by $f = (f_a, f_b)$ with $f_a, f_b : \Omega \to \mathbb{R}$, where $f_a(\omega) = f(a, \omega)$ and $f_b(\omega) = f(b, \omega)$. The space $C_\Omega(X)$ of random continuous functions is thus given by collecting all $f : \Omega \to \mathbb{R}^2$ with $\omega \mapsto \max\{|f_a(\omega)|, |f_b(\omega)|\}$ integrable with respect to P. Since all norms on $\mathbb{R}^2$ are equivalent, $C_\Omega(X)$ can be identified with $L^1(\Omega, \mathscr{F}, P; \mathbb{R}^2)$. Denote by $|\cdot|$ an arbitrary norm on $\mathbb{R}^2$. The space $\mathrm{BL}_\Omega(X)$ of random bounded Lipschitz functions is given by $g : \Omega \to \mathbb{R}^2$ with $\omega \mapsto |g(\omega)|$ bounded P-a.s., so $\mathrm{BL}_\Omega(X)$ can be identified with $L^\infty(\Omega, \mathscr{F}, P; \mathbb{R}^2)$. The space $Pr(X)$ of probability measures consists of the Bernoulli measures, $Pr(X) = \{t\delta_a + (1-t)\delta_b : t \in [0, 1]\}$. In order to be specific, choose $d(\rho, \zeta) = \sup\{\rho(g) - \zeta(g) : g \in \mathrm{BL}(X), 0 \leq g \leq 1\}$ for a metric on $Pr(X)$, metrising the narrow topology. In the present situation this gives for $\rho = t\delta_a + (1-t)\delta_b$ and $\zeta = s\delta_a + (1-s)\delta_b$ with $t, s \in [0, 1]$, and $g_a = g(a)$, $g_b = g(b)$,

$$
\begin{aligned}
d(\rho, \zeta) &= \sup\{tg_a - sg_a + (1-t)g_b - (1-s)g_b : 0 \leq g_a, g_b \leq 1\} \\
&= \sup\{(t-s)(g_a - g_b) : 0 \leq g_a, g_b \leq 1\} \\
&= |t - s|. \qquad\qquad (5.3)
\end{aligned}
$$

Any random measure $\mu \in Pr_\Omega(X)$ is given by

$$
\mu_\omega = t(\omega)\delta_a + (1 - t(\omega))\delta_b
$$

where $\omega \mapsto t(\omega)$ is measurable (since $\omega \mapsto \mu_\omega(\{a\})$ is measurable). Thus $Pr_\Omega(X)$ can be identified with $L = \{t : \Omega \to [0, 1] : t \text{ measurable}\}$, where the identification is given by $(\omega \mapsto \mu_\omega) \mapsto (\omega \mapsto \mu_\omega(\{a\}) = t(\omega))$. The set of Young measures consists of the Dirac measures $\delta_{y(\omega)} = t(\omega)\delta_a + (1 - t(\omega))\delta_b$, given by random variables $y : \Omega \to \{a, b\}$, or equivalently by $t : \Omega \to \{0, 1\}$, making up the set $\{t : \Omega \to \{0, 1\} : t \text{ measurable}\} \subset L$. We then have $L \subset L^p(\Omega, \mathscr{F}, P)$ for every $1 \leq p \leq \infty$. For $p = \infty$, L is homeomorphic to the closed unit ball in $L^\infty(\Omega, \mathscr{F}, P)$ (with homeomorphism given by $f \mapsto 2f - 1$). In particular, L has nonempty interior with respect to the L^∞-topology. This is not the case for $p < \infty$. From (5.3) we have

$$d(\mu_\omega, \nu_\omega) = |t(\omega) - s(\omega)| \qquad (5.4)$$

with μ and ν represented by t and s, respectively.

Let us now compare the diverse topologies.

The narrow topology on $Pr_\Omega(X)$ is the topology making $\mu \mapsto \mu(f)$ continuous for every $f \in C_\Omega(X)$. Thus the narrow topology gives the (trace of the) weak* topology (or (L^∞, L^1)-topology) on $L \subset L^\infty(\Omega, \mathscr{F}, P)$. By Lemma 3.16 the narrow topology on $Pr_\Omega(X)$ is generated by $f \in C_\Omega(X)$ with $0 \leq f \leq 1$ already; so we have proved that the (trace of the) (L^∞, L^∞) topology on L coincides with the (trace of the) weak* topology.

The topology given by the metric

$$\beta(\mu, \nu) = \int_\Omega d(\mu_\omega, \nu_\omega) \, dP(\omega) = \int_\Omega |t(\omega) - s(\omega)| \, dP(\omega)$$

(using (5.4) for the second identity) gives the (trace of the) strong (L^1-) topology on L, considered as a subset of $L^1(\Omega, \mathscr{F}, P)$. Similarly, the topology given by

$$\beta_p(\mu, \nu) = \left(\int_\Omega d(\mu_\omega, \nu_\omega)^p \, dP(\omega) \right)^{1/p} = \left(\int_\Omega |t(\omega) - s(\omega)|^p \, dP(\omega) \right)^{1/p}$$

gives the (trace of the) strong (L^p-) topology on L, considered as a subspace of $L^p(\Omega, \mathscr{F}, P)$, $1 \leq p \leq \infty$. Since L has nonempty interior in $L^\infty(\Omega, \mathscr{F}, P)$, we get that the topology induced by β_∞ on $Pr_\Omega(\{a, b\})$ is not separable, provided $P : \mathscr{F} \to [0, 1]$ takes more than finitely many values (see Behrends [7], Bemerkung 2, p. 163). Recall that the topology induced by β_p is separable for $1 \leq p < \infty$ as soon as $\mathscr{F}$ is countably generated (mod P) by Lemma 5.5.

The Ky Fan metric

$$\varkappa(\mu, \nu) = \inf\{\varepsilon > 0 : P\{\omega : d(\mu_\omega, \nu_\omega) > \varepsilon\} \leq \varepsilon\}$$
$$= \inf\{\varepsilon > 0 : P(|t - s| > \varepsilon) \leq \varepsilon\}$$

metrises convergence in probability on L.

Finally, P-almost sure convergence of μ^n to μ corresponds to P-almost sure convergence of $\omega \mapsto t_n(\omega)$ to $\omega \mapsto t(\omega)$, where t_n and t represent μ^n and μ, respectively. Convergence in law corresponds to convergence of Q_{t_n}, $n \in \mathbb{N}$, where Q_{t_n} stands for the distribution of the real valued random variable t_n.

5.7 Example In the situation discussed above ($X = \{a, b\}$), suppose that the probability space $(\Omega, \mathscr{F}, P)$ is $([0,1], \mathscr{B}([0,1]), \lambda)$, where λ denotes the Lebesgue measure. Let $g_n : \Omega \to [-1, 1]$, $n \in \mathbb{N}$, be the *Rademacher functions*

$$g_n(\omega) = \begin{cases} 1 & \text{for } \dfrac{2k}{2^n} \le \omega < \dfrac{2k+1}{2^n}, \\[2mm] -1 & \text{for } \dfrac{2k+1}{2^n} \le \omega < \dfrac{2k+2}{2^n}, \end{cases}$$

$k \in \mathbb{N}$. Then $(g_n)_{n\in\mathbb{N}}$ is an orthonormal sequence in the Hilbert space $L^2(\Omega, \mathscr{F}, P; \mathbb{R})$ with the usual scalar product

$$\langle h, g \rangle = \int_\Omega hg \, dP.$$

Note that for every $n \neq m$

$$P\{|g_n - g_m| > \eta\} = \frac{1}{2} \tag{5.5}$$

for every $\eta < 2$. Denote by $G = \operatorname{span}\{g_n : n \in \mathbb{N}\}$ the linear span of $(g_n)_{n\in\mathbb{N}}$, and denote by $\mathscr{G} = \sigma\{g_n : n \in \mathbb{N}\}$ the countably generated σ-algebra generated by $(g_n)_{n\in\mathbb{N}}$ (here even without 'mod P'). For any $h \in G$

$$h = \sum_{n\in\mathbb{N}} \langle h, g_n \rangle g_n \qquad \text{and} \qquad \|h\|_2^2 = \sum_{n\in\mathbb{N}} |\langle h, g_n \rangle|^2.$$

Consequently, $\langle h, g_n \rangle$ converges to 0 with $n \to \infty$.
Now put $t_n(\omega) = \frac{1}{2}(g_n(\omega) + 1)$, and define the random measure μ^n by $\mu_\omega^n = t_n(\omega)\delta_a + (1 - t_n(\omega))\delta_b$ for $n \in \mathbb{N}$. Then for any $h : \Omega \to \mathbb{R}^2$, $h = (h_a, h_b)$, with $h_a, h_b \in G$ we have

$$\begin{aligned} \mu^n(h) &= \int_\Omega \left(h_a(\omega)\frac{g_n(\omega)+1}{2} + h_b(\omega)\frac{1 - g_n(\omega)}{2} \right) dP(\omega) \\ &= \frac{1}{2}(Eh_a + Eh_b) + \frac{1}{2}\langle h_a - h_b, g_n \rangle. \end{aligned}$$

Consider the non-random measure $\rho = \frac{1}{2}(\delta_a + \delta_b) \in Pr(\{a, b\})$, interpreting it as a constant random measure. For every $h = (h_a, h_b) : \Omega \to \mathbb{R}^2$ then

$\rho(h) = 1/2 E(h_a + h_b)$, hence convergence of $\langle h_a - h_b, g_n \rangle$ to 0 implies that $\mu^n(h)$ converges to $\rho(h)$ for every h. Consequently, μ^n converges to ρ in the narrow topology of $Pr_{\Omega, \mathcal{G}}(X)$. However, for $n, m \in \mathbb{N}$, $n \neq m$,

$$d(\mu_\omega^n, \mu_\omega^m) = |t_n(\omega) - t_m(\omega)| = \frac{1}{2}|g_n(\omega) - g_m(\omega)|$$

hence $P\{\omega : d(\mu_\omega^n, \mu_\omega^m) > \eta\} = 1/2$ for every $\eta < 1$ by (5.5), whence $\varkappa(\mu^n, \mu^m) = 1/2$ for all $n, m \in \mathbb{N}$, $n \neq m$. Thus μ^n does not converge in the Ky Fan metric, so it does not converge in probability, and, in particular, not P-almost surely. Note that this takes place on the probability space $(\Omega, \mathcal{G}, P)$ with the countably generated σ-algebra $\mathcal{G}$, so that the narrow topology on $Pr_{\Omega, \mathcal{G}}(X)$ is metrisable by Theorem 4.16 in the present situation. Thus convergence in the narrow topology on the random measures $Pr_\Omega(X)$ does not imply convergence in probability, regardless of whether the narrow topology on $Pr_\Omega(X)$ is metrisable or not. Note further that t_n, $n \in \mathbb{N}$, takes only the values 0 and 1, so μ^n is a Young measure for every $n \in \mathbb{N}$, which means that it is a random Dirac measure, supported by the random variable $x_n : \Omega \to \{a, b\}$, $x_n(\omega) = a1_{\{t_n=1\}}(\omega) + b1_{\{t_n=0\}}(\omega)$ (cf. Remark 4.32). The example shows furthermore that μ^n converges to μ in the narrow topology, but that there exist bounded continuous $f \in C(X)$ such that the real random variables $\omega \mapsto \mu_\omega^n(f)$ do not converge in probability. They even do not have any subsequence converging in probability. Choose, e.g., $f = (f_a, f_b)$ with $f_a = 1$ and $f_b = 0$, then $\mu_\omega^n(f) = t_n(\omega)$. Of course, $\omega \mapsto \mu_\omega^n(f)$ does not converge P-almost surely either.

We note that in this example the space of random measures on X is compact in the narrow topology (by Proposition 4.3 together with Theorem 4.4/4.29). However, it is not compact in the topology of convergence in probability, containing a sequence (μ_n) with Ky Fan metric $\varkappa(\mu_n, \mu_m) = 1/2$ for $n \neq m$.

Turning to the question of convergence in law now, let $(\mu^n)_{n \in \mathbb{N}}$ be the sequence of random Dirac measures constructed above, and note that the laws $\mathcal{L}(\mu^n) = \mu^n P$ on the space $Pr(Pr(X))$ of probability measures on $Pr(X)$ are

$$\mathcal{L}(\mu^n) = \frac{1}{2}(\delta_{\delta_a} + \delta_{\delta_b}),$$

and hence they do not depend on $n \in \mathbb{N}$. On the other hand, the law $\mathcal{L}(\rho) = \rho P$ of the limit in the narrow topology, which is the non-random $\rho = \lim \mu^n = 1/2(\delta_a + \delta_b)$, is the Dirac measure

$$\mathcal{L}(\rho) = \delta_{1/2(\delta_a + \delta_b)}.$$

Thus the sequence (μ^n) does converge in law, but its distributions do not converge to $\mathcal{L}(\rho)$, which is the law of the limit of (ν^n) in the narrow topology. To obtain a sequence $(\hat{\mu}^n)_{n\in\mathbb{N}}$ which converges in the narrow topology but not in law, put

$$\hat{\mu}^n = \begin{cases} \mu^k & \text{for } n = 2k+1, \\ \rho & \text{for } k = 2n. \end{cases}$$

Then $(\hat{\mu}^n)_{n\in\mathbb{N}}$ converges to ρ in the narrow topology, but the laws $(\mathcal{L}(\hat{\mu}^n))_{n\in\mathbb{N}}$ jump between $1/2(\delta_{\delta_a} + \delta_{\delta_b})$ and $\delta_{(\delta_a+\delta_b)/2}$.

As noted above, convergence in law to a non-random measure – so convergence of $\mathcal{L}(\nu^n)$ to δ_ζ for some $\zeta \in Pr(X)$ – implies convergence in probability, hence convergence in the Ky Fan metric $\varkappa$, hence convergence in the narrow topology by Proposition 5.4. But here the situation is different: We have convergence of $(\mu^n)_{n\in\mathbb{N}}$ to a non-random ρ in the narrow topology, and we have convergence of $(\mu^n)_{n\in\mathbb{N}}$ in law to $(\delta_{\delta_a} + \delta_{\delta_b})/2$, which is not the law of a non-random measure; then it would have to be a Dirac measure. So this does not give a contradiction, as it might seem on first view.

Note that convergence of μ^n to ρ in the narrow topology does not even imply convergence of $\mu^n(f)$ to $\rho(f)$ in law for $f \in C(X)$. Again $f = (f_a, f_b)$ with $f_a = 1$ and $f_b = 0$ does: $Q_{\mu^n(f)} = Q_{t_n} = (\delta_0 + \delta_1)/2$, whereas $Q_{\rho(f)} = \delta_{1/2}$, where we wrote Q_z for the distribution of an $\mathbb{R}$-valued random variable z.

We had argued above that the topology induced by the Ky Fan metric is strictly stronger than the narrow topology on $Pr_\Omega(X)$ by showing directly that $(\mu^n)_{n\in\mathbb{N}}$ is not a Cauchy sequence in the Ky Fan metric. Of course the assertion on the topologies already follows from the fact that convergence in the narrow topology does not imply convergence in law (because convergence in the Ky Fan metric is equivalent to convergence in probability, which implies convergence in law by (5.2)). We have included the direct argument above both because it seems clearer, and because it does not use (5.2). The example of the Rademacher functions appears also in Valadier [37], Example 1, p. 156.

5.8 Remark (i) Every nontrivial Polish space X contains two distinct points a and b, and the set of all random measures supported by $\{a, b\}$ is a closed subset of $Pr_\Omega(X)$ in the narrow topology by Corollary 3.18. Consequently, Example 5.7 shows that in general convergence of a sequence $(\mu^n)_{n\in\mathbb{N}}$ to $\mu \in Pr_\Omega(X)$ in the narrow topology does not even imply convergence in law of the sequence of real valued random variables $(\omega \mapsto \int f(x)\,d\mu^n_\omega(x))_{n\in\mathbb{N}}$ for $f \in C(X)$ – not to speak of convergence in probability or P-almost sure convergence.

(ii) Example 5.7 shows that in general the map

$$\mathcal{L} : Pr_\Omega(X) \;\to\; Pr\big(Pr(X)\big),$$
$$\mu \;\mapsto\; \mathcal{L}(\mu) = \mu P,$$

is not continuous with respect to the respective narrow topologies on the spaces $Pr_\Omega(X)$ and $Pr\big(Pr(X)\big)$. In fact, this carries over to any Hausdorff topology on $Pr\big(Pr(X)\big)$. Recall that $\mathcal{L}$ is continuous with respect to the (stronger) topology induced by the Ky Fan metric $\varkappa$ on $Pr_\Omega(X)$ and the narrow topology on $Pr\big(Pr(X)\big)$ by (5.2).

The law of an invariant measure for a random dynamical system (RDS) is sometimes addressed to as a *statistical equilibrium*. It is remarkable that the narrow topology, which is the basic topology on random measures for RDS, is not strong enough to allow continuous dependence of the statistical equilibrium on the random measure under consideration. However, there is an important limiting procedure for invariant measures, which gives P-almost sure convergence, hence also convergence in the Ky Fan metric, and thus also both convergence of the law as well as convergence in the narrow topology (see, e.g., Arnold [1], Theorem 1.7.2, p. 37).

(iii) Example 5.7 also shows that tightness does not imply relative compactness in the (metrisable) $\varkappa$-topology on $Pr_\Omega(X)$. In fact, $Pr_\Omega(\{a, b\})$ is tight (as $Pr_\Omega(X)$ is always compact, hence tight, for X compact), but $\varkappa(\mu^n, \mu^m) = 1/2$ for $n \neq m$, whence $(\mu^n)_{n \in \mathbb{N}}$ does not have convergent subsequences in the topology induced by $\varkappa$.

We summarise our considerations on the different topologies on $Pr_\Omega(X)$ by a diagram showing which convergence implies which other convergences. To read the diagram, suppose that $(\mu^n)_{n \in \mathbb{N}}$ is a sequence in $Pr_\Omega(X)$. Then every arrow stands for 'implies convergence in'

$$\beta_p \longrightarrow \beta_1 \qquad\qquad \text{in } \mathcal{L}$$
$$\beta_\infty \Big\langle \qquad\qquad\qquad \Big\rangle \varkappa \Big\langle$$
$$P\text{-a.s.} \qquad\qquad\qquad \text{in the narrow topology}$$

and both the lower and upper rows do not imply the respective opposite upper and lower row. If the metric d on $Pr(X)$ is chosen to be bounded, then the arrows between β_p, β_1 and $\varkappa$ also go in the other direction for $p < \infty$ by Lemma 5.2.

Finally, suppose that $(\mu^n)_{n \in \mathbb{N}} \subset Pr_\Omega(X)$ is a sequence of random measures. We want to discuss the question whether convergence of $(\omega \mapsto \mu^n_\omega(f))_{n \in \mathbb{N}}$ in probability for every $f \in C(X)$, the space of bounded continuous functions on X, implies convergence of $(\mu^n)_{n \in \mathbb{N}}$ in the narrow topology. This condition means that, for every $f \in C(X)$, the sequence $\mu^n(f)$ converges in the Ky Fan metric on the space of $\mathbb{R}$-valued random variables. This is weaker than convergence of $(\mu^n)_{n \in \mathbb{N}}$ in the Ky Fan metric related to the $Pr(X)$-valued random variables (which is known to imply convergence of $(\mu^n)_{n \in \mathbb{N}}$ by Proposition 5.4).

5.9 Lemma *Suppose that $(\mu^n)_{n \in \mathbb{N}} \subset Pr_\Omega(X)$ is a sequence of random measures such that for every $f \in C(X)$ the sequence of $\mathbb{R}$-valued random variables $(\omega \mapsto \mu_\omega^n(f))_{n \in \mathbb{N}}$ converges in probability. Then there exists $\mu \in Pr_\Omega(X)$ such that $\lim_{n \to \infty} \mu^n = \mu$ in the narrow topology.*

PROOF For every $f \in C(X)$ we have $\nu_\omega(f) \leq \sup_x |f(x)|$ for all $\nu \in Pr_\Omega(X)$, hence the sequence $(\omega \mapsto |\mu_\omega^n(f)|)_{n \in \mathbb{N}}$ is uniformly integrable. Therefore, $\int \mu_\omega^n(f) \, dP(\omega)$ converges with $n \to \infty$. But $\int \mu_\omega^n(f) \, dP(\omega) = \pi_X \mu^n(f)$ by Lemma 3.22, where $\pi_X \mu^n$ is the marginal of μ^n on X. Denoting $\rho_n = \pi_X \mu^n$, $n \in \mathbb{N}$, we thus have convergence of $(\rho_n(f))_{n \in \mathbb{N}}$ for every $f \in C(X)$, so the sequence $(\rho_n)_{n \in \mathbb{N}}$ converges in $Pr(X)$. Denote the limit by $\rho \in Pr(X)$. In particular, $\{\rho_n : n \in \mathbb{N}\}$ is relatively compact, hence tight. Since $\pi_X \{\mu^n : n \in \mathbb{N}\} = \{\rho_n : n \in \mathbb{N}\}$, also $\{\mu^n : n \in \mathbb{N}\}$ is tight (see Definition 4.2), and hence it is relatively sequentially compact in the narrow topology by Theorem 4.4. Consequently, $(\mu^n)_{n \in \mathbb{N}}$ has a subsequence which converges in the narrow topology to some $\mu \in Pr_\Omega(X)$ (which necessarily has marginal $\pi_X \mu = \rho$ by continuity of π_X, see Corollary 4.27). We now claim that $(\mu^n)_{n \in \mathbb{N}}$ converges to μ. Suppose that there is another subsequence converging to some $\hat{\mu} \in Pr_\Omega(X)$, say. Convergence of $(\omega \mapsto \mu_\omega^n(f))_{n \in \mathbb{N}}$ in probability implies convergence of $(\omega \mapsto \mu_\omega^n(f) 1_F(\omega))_{n \in \mathbb{N}}$ in probability for every $f \in C(X)$ and $F \in \mathscr{F}$. Furthermore, $(\omega \mapsto \mu_\omega^n(f) 1_F(\omega))_{n \in \mathbb{N}}$ is uniformly integrable, hence

$$\int_\Omega 1_F(\omega) \int_X f(x) \, d\mu_\omega^n(x) \, dP(\omega) = \mu^n(f 1_F)$$

converges for every $f \in C(X)$ and $F \in \mathscr{F}$. But $(x,\omega) \mapsto f(x) 1_F(\omega) \in C_\Omega(X)$, so that convergence along the corresponding subsequences of $(\mu^n)_{n \in \mathbb{N}}$ in the narrow topology implies $\mu(f 1_F) = \hat{\mu}(f 1_F)$ for every $f \in C(X)$ and $F \in \mathscr{F}$. But this is sufficient to conclude $\mu = \hat{\mu}$ by Lemma 3.14. Therefore $(\mu^n)_{n \in \mathbb{N}}$ converges to μ in the narrow topology. $\qquad \square$

Chapter 6

Invariant Measures and Some Ergodic Theory for Random Dynamical Systems

In the present chapter the results obtained before are going to be applied to random dynamical systems (RDS). We will give only the abstract characterisation of RDS. For a comprehensive and thorough description of the generation of RDS from random and stochastic differential equations in finite dimensional spaces we refer to Arnold [1], in particular to Chapter 2. Concerning the generation of RDS from infinite-dimensional problems, in particular from stochastic parabolic stochastic partial differential equations on bounded domains, see Flandoli [23].

The results we are interested here are assertions about existence of invariant measures and of invariant Markov measures for RDS on Polish spaces, as well as a characterisation of the convergence of time means in terms of integrals over ergodic invariant measures. These results are the ones which need the topological prerequisites obtained in previous chapters.

For a large variety of further topics in the theory of RDS – the multiplicative ergodic theorem, invariant manifolds, normal forms, bifurcation theory, etc. – we refer to Arnold [1] and references given there. We note that for finite-dimensional state spaces or, more generally, for locally compact Hausdorff spaces with a countable base (LCCB), existence of invariant measures for RDS can be obtained with considerably less effort. It is only for state spaces which are not locally compact (but only Polish) that we need the Prohorov theory for random measures as developed in the previous chapters.

Random Dynamical Systems

Suppose that Y is a set, and $\vartheta : Y \to Y$ is a map. A set $B \subset Y$ is said to be

- *invariant under* ϑ or ϑ-*invariant* if $\vartheta B \subset B$ (or, equivalently, $B \subset \vartheta^{-1}B$),

- *strictly invariant under* ϑ or *strictly* ϑ-*invariant* if $\vartheta B = B$ (which is not equivalent to $B = \vartheta^{-1}B$).

If $\{\vartheta_\alpha : Y \to Y : \alpha \in A\}$ is a family of maps then $B \subset Y$ is said to be $(\vartheta_\alpha)_\alpha$-*invariant* or *strictly* $(\vartheta_\alpha)_\alpha$-*invariant*, respectively, if B is ϑ_α-invariant or strictly ϑ_α-invariant, respectively, for all $\alpha \in A$.

Suppose that $(Y, \mathcal{Y})$ is a measurable space. Recall that a map $f : Y \to Y$ is said to be *measurable* if $f^{-1}B \in \mathcal{Y}$ for every $B \in \mathcal{Y}$. It is said to be *bimeasurable* if it is measurable and invertible, and if f^{-1} is also measurable. If f is bimeasurable, then $f^{-1}\mathcal{Y} = \mathcal{Y} = f\mathcal{Y}$.

Suppose that $(Y, \mathcal{Y}, m)$ is a measure space. A measurable map $\vartheta : Y \to Y$ is said to be m-*preserving* or *measure preserving* (if there is no ambiguity concerning the measure) if $m(\vartheta^{-1}F) = m(F)$ for all $F \in \mathcal{F}$, i.e., if $\vartheta m = m$. Under the same conditions m is said to be ϑ-*invariant* or just *invariant* (if there is no ambiguity concerning the transformation).

Suppose that $\{\vartheta_\alpha : \alpha \in A\}$ is a family of measurable transformations of $(Y, \mathcal{Y})$. Suppose further that m is a probability measure on $(Y, \mathcal{Y})$. The pair $(m, \{\vartheta_\alpha : \alpha \in A\})$ is said to be *ergodic* if $m(\vartheta_\alpha^{-1}B \bigtriangleup B) = 0$ for all $\alpha \in A$ implies $m(B) = 0$ or $m(B) = 1$. If there is no ambiguity about m we say that $\{\vartheta_\alpha : \alpha \in A\}$ is ergodic, and if there is no ambiguity about $\{\vartheta_\alpha : \alpha \in A\}$ we say that m is ergodic. Note that this notion of ergodicity does not assume every single ϑ_α to be ergodic.

Let T be either $\mathbb{N}$, $\mathbb{Z}$, $\mathbb{R}^+$ or $\mathbb{R}$, understood with their respective natural topologies (and Borel σ-algebras). We imagine T as 'time'. We speak of *one-sided time* if T is either $\mathbb{N}$ or $\mathbb{R}^+$, and of *two-sided time* if T is either $\mathbb{Z}$ or $\mathbb{R}$. We further speak of *discrete time* if T is either $\mathbb{N}$ or $\mathbb{Z}$, and of *continuous time* if T is either $\mathbb{R}^+$ or $\mathbb{R}$.

Suppose that $(\Omega, \mathcal{F}, P)$ is a probability space. Let $\vartheta : T \times \Omega \to \Omega$ be a measurable map, and denote by $\vartheta_t(\cdot) = \vartheta(t, \cdot)$ the map from Ω to itself obtained by fixing $t \in T$. Suppose furthermore that the family $(\vartheta_t)_{t \in T}$ satisfies

(i) ϑ_t preserves the measure P for each $t \in T$,

(ii) $\vartheta_{t+s} = \vartheta_t \circ \vartheta_s$ for all $t, s \in T$, and $\vartheta_0 = \text{id}$.

The set-up $(\Omega, \mathscr{F}, P; (\vartheta_t)_{t \in T})$ is one of the two basic constituents of an RDS. The family $(\vartheta_t)_{t \in T}$ is sometimes called a *flow* (*of measure preserving transformations*) *on* $(\Omega, \mathscr{F}, P)$ (which is misuse of notation for T one-sided).

6.1 Remark (i) If T is two-sided then ϑ_t is invertible and bimeasurable for each $t \in T$.

(ii) If T is discrete then ϑ consists of the iterates of one single map, $\vartheta_n = \vartheta_1^n$, $n \in \mathbb{N}$ or $n \in \mathbb{Z}$.

6.2 Definition (Random Dynamical System) Suppose that T_{base} is either of $\mathbb{R}$, $\mathbb{R}^+$, $\mathbb{Z}$ or $\mathbb{N}$, and that $(\vartheta_t)_{t \in T_{\text{base}}}$ is a flow of measure preserving transformations on $(\Omega, \mathscr{F}, P)$. Suppose further that T is either of $\mathbb{R}$, $\mathbb{R}^+$, $\mathbb{Z}$ or $\mathbb{N}$, and $T \subset T_{\text{base}}$. An *RDS* with time T on a Polish space X with Borel σ-algebra $\mathscr{B}$ over $(\Omega, \mathscr{F}, P; (\vartheta_t)_{t \in T_{\text{base}}})$ is a measurable map

$$\varphi : T \times X \times \Omega \;\to\; X$$
$$(t, x, \omega) \;\mapsto\; \varphi(t, \omega)x$$

such that $\varphi(0, \omega) = \text{id}$ (identity on X), and such that for every $s \in T$ there exists $N_s \in \mathscr{F}$ with $P(N_s) = 0$ such that

$$\varphi(t + s, \omega) = \varphi(t, \vartheta_s\omega) \circ \varphi(s, \omega) \tag{6.1}$$

for all $t \in T$ and for all $\omega \in \Omega \setminus N_s$, where $\circ$ means composition. A family of maps $\varphi(t, \omega)$ satisfying (6.1) is called a (*crude*) *cocycle*. In case (6.1) holds for all $t, s \in T$ and for all ω from a measurable $(\vartheta_t)_{t \in T}$-invariant set of full measure, the cocycle φ is said to be *perfect*. Equation (6.1) is the *cocycle property*.

6.3 Remark The definition of an RDS usually assumes a perfect cocycle, where (6.1) is assumed to hold for all $s, t \in T$ and all $\omega \in \Omega$ (which is essentially equivalent to have it for ω from a $(\vartheta_t)_{t \in T}$-invariant subset of full measure). The problem comes from the dependence of the nullset N_s on s. Since it is the weaker 'crude' form which is really needed in the following, we here allow also for crude RDS.

6.4 Remark (i) If the system time T is two-sided then $\varphi(t, \omega)$ is invertible for P-almost all ω for every $t \in T$, and $\varphi(t, \omega)^{-1} = \varphi(-t, \vartheta_t\omega)$, hence $\varphi(t, \omega)$ is bimeasurable for P-almost all ω, for every $t \in T$.

(ii) For T discrete an RDS is given by the map $\varphi : X \times \Omega \to X$, $\varphi(\omega)x = \varphi(1, \omega)x$ already.

6.5 Remark The time T_{base} of the base system (the 'base time') can be strictly larger than the system time T. The most interesting case is the one where T_{base} is two-sided, whereas T is one-sided. For continuous time this occurs, e.g., if φ comes from an infinite-dimensional problem from the applications, which often do not act injectively on initial conditions. These systems thus cannot have solutions for time going backward. Another example for a natural occurrence of one-sided time is given by random differential equations which explode when time goes backwards. This can be handled by introducing 'local RDS' (see Arnold [1], Chapter 1.2, pp. 11–15). For discrete time, non-invertibility of $\varphi(1,\omega)$ with positive probability implies that T must be one-sided by Remark 6.4 (i).

If the base time T_{base} is not specified in the following we will write (ϑ_t) for $(\vartheta_t)_{t\in T}$, and the corresponding assertions refer to any admissible choice of $T_{\text{base}} \supset T$.

An RDS is said to be *continuous* or *differentiable*, respectively, if $\varphi(t,\omega)$: $X \to X$ is continuous or differentiable, respectively, for all $t \in T$ and all $\omega \in \Omega$ (or for all ω outside a P-nullset not depending on t). For a continuous discrete time RDS on a separable metric space the assumption on joint measurability of $(x,\omega) \mapsto \varphi(\omega)x$ is automatically satisfied (see Lemma 1.1). If T is two-sided then for every $t \in T$ the map $\varphi(t,\omega)$: $X \to X$ is a homeomorphism for P-almost all ω (where the exceptional nullset can depend on t), cf. Remark 6.4 (i). Note that we do not assume continuity in time, so $t \mapsto \varphi(t,\omega)x$ need not be continuous.

6.6 Definition Suppose that φ is an RDS on X. The map

$$\Theta : T \times X \times \Omega \;\to\; X \times \Omega,$$
$$(t,x,\omega) \;\mapsto\; (\varphi(t,\omega)x, \vartheta_t\omega)$$

is said to be the *skew product* induced by φ.

Note that Θ is measurable by the assumptions on φ and ϑ. Given $t \in T$, write $\Theta_t(x,\omega) = \Theta(t,x,\omega)$ for $(x,\omega) \in X \times \Omega$. Then $\Theta_{t+s} = \Theta_t \circ \Theta_s$ for all $t,s \in T$, and $\Theta_0 = \text{id}$ on $X \times \Omega$. The family Θ_t, $t \in T$, is called the *skew product (semi-) flow* induced by φ.

Suppose that φ is an RDS with time T, on a Polish space X with Borel σ-algebra $\mathscr{B}$, over (ϑ_t) on $(\Omega, \mathscr{F}, P)$. Denote by $(\Theta_t)_{t\in T}$ the skew product flow induced by φ,

$$\Theta_t : X \times \Omega \;\to\; X \times \Omega,$$
$$(x,\omega) \;\mapsto\; (\varphi(t,\omega)x, \vartheta_t\omega).$$

For every $t \in T$ the skew product map Θ_t acts on functions on $X \times \Omega$ in the usual way, $\Theta_t f(x, \omega) = (f \circ \Theta_t)(x, \omega) = f(\varphi(t, \omega)x, \vartheta_t \omega)$ for a function f on $X \times \Omega$. It also acts on measures on $X \times \Omega$ in the usual way, $\Theta_t \mu(A) = \mu(\Theta_t^{-1} A)$ for $A \in \mathscr{B} \otimes \mathscr{F}$. We refrain from introducing a new notation for the action of Θ_t on either functions or measures. The actions of Θ_t on functions and on measures are related by

$$\int_{X \times \Omega} \Theta_t h \, d\mu = \int_{X \times \Omega} h \, d(\Theta_t \mu) \tag{6.2}$$

for all measures μ on $\mathscr{B} \otimes \mathscr{F}$ and for all measurable $h : X \times \Omega \to \mathbb{R}$ which are μ-integrable (Gänssler and Stute [24], Satz 1.10.4, p. 54).

6.7 Lemma *Suppose that φ is a continuous RDS. Then, for each $t \in T$,*

(i) Θ_t *maps $C_\Omega(X)$ to itself, and Θ_t is continuous on $C_\Omega(X)$ (with respect to the topology induced by the norm $|\cdot|_\infty$ from Definition 3.9),*

(ii) Θ_t *maps $Pr_\Omega(X)$ to itself, and Θ_t is continuous on $Pr_\Omega(X)$ (with respect to the narrow topology).*

PROOF (i) For every $t \in T$ and for every $\omega \in \Omega$ the map $x \mapsto \Theta_t(x, \omega)$ is continuous, and for every $x \in X$ the map $\omega \mapsto \Theta_t(x, \omega)$ is measurable. For $f \in C_\Omega(X)$ thus $x \mapsto \Theta_t f(x, \omega) = f(\varphi(t, \omega)x, \vartheta_t \omega)$ is continuous for each $\omega \in \Omega$ (regardless of the version of f), and $\omega \mapsto \Theta_t f(x, \omega)$ is measurable for each $x \in X$ (by Lemma 1.1). Furthermore, if f and f' are versions of each other, then

$$\{\omega : \Theta_t f(\cdot, \omega) \neq \Theta_t f'(\cdot, \omega)\} \subset \{\omega : f(\cdot, \vartheta_t \omega) \neq f'(\cdot, \vartheta_t \omega)\}$$

is measurable (using separability of X) and contained in a P-nullset. Finally,

$$
\begin{aligned}
|\Theta_t f - \Theta_t g|_\infty &= \int \sup_{x \in X} |f(\varphi(t, \omega)x, \vartheta_t \omega) - g(\varphi(t, \omega)x, \vartheta_t \omega)| \, dP(\omega) \\
&\leq \int \sup_{y \in X} |f(y, \vartheta_t \omega) - g(y, \vartheta_t \omega)| \, dP(\omega) = |f - g|_\infty.
\end{aligned}
$$

(ii) Since $\mu \mapsto \Theta_t \mu(f) = \mu(\Theta_t f)$ for every $f \in C_\Omega(X)$ by (6.2), continuity of $\mu \mapsto \Theta_t \mu$ is immediate from (i). $\quad\square$

6.8 Remark Continuity of $\varphi(t)$ on X as well as continuity of Θ_t on either $C_\Omega(X)$ or $Pr_\Omega(X)$ is independent of the choice of a metric on X. On $C_\Omega(X)$ the map Θ_t is always Lipschitz with respect to $|\cdot|_\infty$. But note that Θ_t in general does not map the space $BL_\Omega(X)$ of random Lipschitz functions to itself.

Invariant Measures

6.9 Definition (INVARIANT MEASURE, INVARIANT SET) Let φ be a continuous RDS.

(i) A random measure μ is said to be an *invariant measure for* φ if it is invariant for the skew product flow induced by φ, i.e., if $\Theta_t \mu = \mu$ for all $t \in T$. It is said to be an *ergodic invariant measure for* φ if it is invariant for φ and if μ is ergodic for $\{\Theta_t : t \in T\}$.

(ii) A measurable set $\omega \mapsto B(\omega)$ is said to be *(forward) invariant for* φ if

$$\varphi(t, \omega) B(\omega) \subset B(\vartheta_t \omega)$$

P-a.s., for every $t > 0$.

(iii) A measurable set $\omega \mapsto B(\omega)$ is said to be *strictly invariant for* φ if

$$\varphi(t, \omega) B(\omega) = B(\vartheta_t \omega)$$

P-a.s., for every $t \in T$.

6.10 Remark (i) For a random measure μ to be invariant for an RDS φ it suffices to have $\Theta_t \mu = \mu$ for $t > 0$. Indeed, if T contains only nonnegative times, this is only the definition itself. In case $-t \in T$ for some $t > 0$ this follows from $\Theta_{-t} \mu = \Theta_{-t}(\Theta_t \mu) = \Theta_{-t+t} \mu = \mu$.

(ii) Likewise, for a measurable set to be strictly invariant it suffices to have $\varphi(t, \omega) B(\omega) = B(\vartheta_t \omega)$ P-a.s. for $t > 0$.

(iii) Ergodic invariant measures for an RDS cannot exist if the base flow (ϑ_t) is not ergodic with respect to P.

The main result on the existence of invariant measures can be proved quite easily now. We will provide two formulations of this result. One refers to a general fixed point theorem (Markov–Kakutani; for discrete time Brouwer suffices), which makes it short and elegant. The other one constructs an invariant measure more directly. In the following we will consider the space $Pr_\Omega(X)$ of random probability measures as a subspace of the space of random finite signed measures $\mathcal{M}_\Omega(X)$ without further mentioning. The narrow topology extends to $\mathcal{M}_\Omega(X)$ in the obvious way, making $\mathcal{M}_\Omega(X)$ a locally convex topological vector space. Compare Definition 3.15; every element of the neighbourhood basis given by sets of the form (3.5) is convex.

6.11 Theorem (EXISTENCE OF AN INVARIANT MEASURE VIA THE MARKOV–KAKUTANI FIXED POINT THEOREM) *Suppose that φ is a continuous RDS on a Polish space X. Suppose further that $\Gamma \subset Pr_\Omega(X)$ is a closed, tight, and convex set of random measures such that $\Theta_t \Gamma \subset \Gamma$ for all $t > 0$. If $\Gamma \neq \emptyset$ then it contains an invariant measure for φ.*

PROOF   For each $t \in T$, Θ_t is convex and continuous with respect to the narrow topology by Lemma 6.7 (ii). By Theorem 4.4 Γ is compact in the narrow topology. Furthermore, Θ_t and Θ_s commute for any $s, t \in T$; see the remark after Definition 6.6. It now follows from the Markov–Kakutani fixed point theorem (Dunford and Schwartz [20], Theorem V.10.5, p. 456) that there exists $\mu \in \Gamma$ with $\Theta_t \mu = \mu$ for all $t \geq 0$, and hence the assertion follows in view of Remark 6.10.   □

Now we prove existence of an invariant measure under the same conditions as those of Theorem 6.11, but here by constructing the invariant measure over time means. This type of argument is often addressed to as 'Krylov–Bogolyubov procedure'.

The integral in the following theorem is to be understood as a sum in case T is discrete.

6.12 Theorem (EXISTENCE OF AN INVARIANT MEASURE VIA A KRYLOV–BOGOLYUBOV TYPE ARGUMENT) *Suppose that φ is a continuous RDS on a Polish space X, and that $\emptyset \neq \Gamma \subset Pr_\Omega(X)$ is a closed, tight, and convex set of random measures such that $\Theta_t \Gamma \subset \Gamma$ for all $t > 0$. Let σ^n, $n \in \mathbb{N}$, be an arbitrary sequence in Γ, and let $t_n : \Omega \to T$ be an increasing sequence of nonnegative random times such that $\limsup E(t_n^{-1}) = 0$ for $n \to \infty$. Then the sequence $(\gamma^n)_n \subset \Gamma$, given by*

$$\omega \mapsto \gamma_\omega^n = \frac{1}{t_n(\omega)} \int_0^{t_n(\omega)} \Theta_s \sigma_\omega^n \, ds, \tag{6.3}$$

has a convergent subsequence. Furthermore, every convergent subsequence of $(\gamma^n)_n$ converges to an invariant measure for φ.

PROOF   By Theorem 4.4, Γ is sequentially compact, and since $\gamma^n \in \Gamma$, $n \in \mathbb{N}$ (by convexity and Θ_t-invariance of Γ), $(\gamma^n)_{n \in \mathbb{N}}$ has a convergent subsequence. We denote this subsequence by $(\gamma^n)_{n \in \mathbb{N}}$ again, and denote its limit by γ. Since the sequence (t_n) is increasing, $\limsup_n E(t_n^{-1}) = \lim_n E(t_n^{-1}) = 0$ along any subsequence. Thus for any $t \in T$ the Borel–Cantelli Lemma yields $\lim_n P\{\omega : t_n(\omega) < t\} = 0$. Fix $t \in T$, $t > 0$. We want to show that γ is invariant under Θ_t. By Lemma 3.14 it suffices to prove that $\Theta_t \gamma(f) = \gamma(f)$ for every $f \in C_\Omega(X)$ with $0 \leq f \leq 1$. For any such $f \in C_\Omega(X)$

$$|\Theta_t \gamma(f) - \gamma(f)|$$

$$= \left| \int_{X \times \Omega} \Theta_t f \, d\gamma - \int_{X \times \Omega} f \, d\gamma \right| = \lim_{n \to \infty} \left| \int_{X \times \Omega} (\Theta_t f - f) \, d\gamma^n \right|$$

$$= \lim_{n \to \infty} \left| \int_\Omega \frac{1}{t_n(\omega)} \int_X \int_0^{t_n(\omega)} (\Theta_{t+s} f(x,\omega) - \Theta_s f(x,\omega)) \, ds \, d\sigma_\omega^n(x) \, dP(\omega) \right|$$

$$\leq \limsup_{n \to \infty} \left[\int_{\{\omega : t_n(\omega) \geq t\}} \frac{1}{t_n(\omega)} \left| \int_t^{t+t_n(\omega)} \left(\int_X \Theta_s f(x,\omega) \, d\sigma_\omega^n(x) \right) ds \right. \right.$$

$$\left. \left. - \int_0^{t_n(\omega)} \left(\int_X \Theta_s f(x,\omega) \, d\sigma_\omega^n(x) \right) ds \right| dP(\omega) + 2P\{\omega : t_n(\omega) < t\} \right]$$

$$\leq \limsup_{n \to \infty} \left[\int_\Omega \frac{1}{t_n(\omega)} \left(\int_{t_n(\omega)}^{t+t_n(\omega)} \sup_{x \in X} |\Theta_s f(x,\omega)| \, ds \right. \right.$$

$$\left. \left. + \int_0^t \sup_{x \in X} |\Theta_s f(x,\omega)| \, ds \right) dP(\omega) + 2P\{\omega : t_n(\omega) < t\} \right]$$

$$\leq \limsup_{n \to \infty} 2t \int_\Omega t_n^{-1} dP + 2 \lim_{n \to \infty} P\{\omega : t_n(\omega) < t\}$$

$$= 2t \limsup_{n \to \infty} E(t_n^{-1}) = 0,$$

where we have used $\sup_{x \in X} |\Theta_s f(x,\omega)| \leq 1$ for all $s \in T$. Consequently, γ is invariant for φ.   □

We note that the conditions of Theorems 6.11 and 6.12 are satisfied, in particular, for $\Gamma = Pr_\Omega(X)$ in case the state space X is compact. However, for many interesting and relevant RDS the assumption of a compact state space is not appropriate. In this case $Pr_\Omega(X)$ is not tight, and verification of tightness of a given set $\Gamma \subset Pr_\Omega(X)$ is not completely trivial. Fortunately, for Γ being the set of all random measures supported by an invariant compact random set tightness can be verified.

6.13 Corollary (Existence of an Invariant Measure Supported by an Invariant Compact Random Set) *Suppose that $\omega \mapsto K(\omega)$ is a compact random set which is forward invariant for φ, and that $K(\omega) \neq \emptyset$ for P-almost all ω. Then there exist invariant measures for φ which are supported by K (i.e., $\mu_\omega(K(\omega)) = 1$ P-a.s.). In particular, for an arbitrary sequence of random measures σ^n with $\sigma_\omega^n(K(\omega)) = 1$ P-a.s., define γ^n by (6.3), $n \in \mathbb{N}$. Then the sequence (γ^n) has convergent subsequences, each of which converges to an invariant measure for φ supported by K.*

PROOF The set

$$\Gamma = \{\mu \in Pr_\Omega(X) : \mu_\omega(K(\omega)) = 1 \quad P\text{-a.s.}\}$$

is tight (Proposition 4.3), convex, and $\Theta_t \Gamma \subset \Gamma$ for $t > 0$ by invariance of K. Also $\Gamma \neq \emptyset$, since by Theorem 2.6 there exists a measurable selection $\omega \mapsto k(\omega) \in K(\omega)$, so $\delta_{k(\omega)}(K(\omega)) = 1$ for P-almost all ω, whence $(\omega \mapsto \delta_{k(\omega)}) \in \Gamma$. It remains to show that Γ is closed. But if μ^α is a net in Γ converging to some μ in $Pr_\Omega(X)$, then

$$1 = \limsup_\alpha \mu^\alpha(K) \leq \mu(K) = \int_\Omega \mu_\omega(K(\omega)) \, dP(\omega)$$

by Theorem 3.17. Hence $\mu \in \Gamma$, so Γ is closed. The assertion of the Corollary therefore follows from Theorem 6.12. $\qquad\square$

Invariant Markov Measures

We now consider the particular case where the base time set T_{base} is two-sided, whereas the system time $T \subset T_{\text{base}}$ can be either one- or two-sided. For each $t \in T$ put $\sigma(\omega \mapsto \varphi(t, \omega)) = \sigma\{\omega \mapsto \varphi(t, \omega)x : x \in X\} \subset \mathscr{F}$. For a continuous RDS φ this gives a separable σ-algebra, since for any countable dense subset $D \subset X$ we have $\sigma(\varphi(t, \cdot)) = \sigma\{\omega \mapsto \varphi(t, \omega)x : x \in D\}$.

6.14 Definition (MARKOV MEASURES) For an RDS φ with two-sided base time T_{base}, the σ-algebra

$$\mathscr{F}_{\leq 0} = \sigma\{\omega \mapsto \varphi(t, \vartheta_{-s}\omega)x : x \in X, \ 0 \leq t \leq s\}$$

is said to be *the past of φ* or *the past of the system*. Elements of the set $Pr_{\Omega, \mathscr{F}_{\leq 0}}(X)$, which consists of random measures that are measurable with respect to the past of φ, are called *Markov measures*.

6.15 Remark (i) For any $\tau \geq 0$ we get

$$(\vartheta_{-\tau})^{-1}\mathscr{F}_{\leq 0} = \sigma\{\omega \mapsto \varphi(t, \vartheta_{-s-\tau}\omega) : x \in X, \ 0 \leq t \leq s\} \subset \mathscr{F}_{\leq 0}, \qquad (6.4)$$

or equivalently $\vartheta_{-\tau} : (\Omega, \mathscr{F}_{\leq 0}) \to (\Omega, \mathscr{F}_{\leq 0})$, which can be interpreted as 'invariance of the past of the system against going back in the time of the world'.

(ii) In case the system time T is two-sided, $\varphi(t, \cdot)$ is a homeomorphism for all $t \in T$, and $\mathscr{F}_{\leq 0} = \sigma\{\varphi(-t, \omega) : t \geq 0\}$ (using $\varphi(t, \vartheta_s\omega) = \varphi(t+s, \omega)\circ\varphi(s, \omega)^{-1}$ for all $s, t \in T$).

(iii) Markov measures have been introduced and investigated by the author (see [11] and [12]).

(iv) Similarly to the past one can introduce the future $\mathscr{F}_{\geq 0} = \sigma\{\omega \mapsto \varphi(t, \vartheta_s\omega) : 0 \leq s, t\}$ of φ. Then $\vartheta_\tau^{-1}\mathscr{F}_{\geq 0} \subset \mathscr{F}_{\geq 0}$ for $\tau \geq 0$. Note that Θ_t acts quite differently on the spaces $Pr_\Omega(X)$ and $C_\Omega(X)$, what reference to past and future is concerned. In the particular case where the RDS is induced by a stochastic flow, past and future are independent. This makes things easy for stochastic flows, see Crauel [11].

6.16 Lemma *Suppose that T_{base} is two-sided. Then, for every $t \geq 0$ and every $\mu \in Pr_\Omega(X)$, the random measure $(\Theta_t\mu) : \Omega \to Pr(X)$ satisfies*

$$(\Theta_t\mu)_\omega = \varphi(t, \vartheta_{-t}\omega)\mu_{\vartheta_{-t}\omega} \tag{6.5}$$

P-a.s. (where the right-hand side is the composition of the maps $\omega \mapsto \vartheta_{-t}\omega$ and $\omega \mapsto \varphi(t, \omega)\mu_\omega$, well defined since T_{base} is two-sided). In particular, for $t \geq 0$

$$\Theta_t : Pr_{\Omega, \mathscr{F}_{\leq 0}}(X) \to Pr_{\Omega, \mathscr{F}_{\leq 0}}(X), \tag{6.6}$$

so Markov measures are mapped to Markov measures under Θ_t for nonnegative times $t \in T$.

PROOF For any $s \in T_{\text{base}}$ and any $\mu \in Pr_\Omega(X)$ also $\omega \mapsto \mu_{\vartheta_s\omega}$ is a random measure, see Definition 3.1. For any $t \in T$ further $\omega \mapsto \varphi(t, \omega)\mu_\omega$ defines a random measure. In fact, $\varphi(t, \omega)\mu_\omega \in Pr(X)$ for every $\omega \in \Omega$, and hence by Remark 3.20 (i) it suffices to establish that $\omega \mapsto (\varphi(t, \omega)\mu_\omega)(f)$ is measurable for every $f \in C(X)$. But $\omega \mapsto (\varphi(t, \omega)\mu_\omega)(f) = \int_X f(\varphi(t, \omega)x) \, d\mu_\omega(x)$, and since $(x, \omega) \mapsto f(\varphi(t, \omega)x)$ is bounded and measurable, measurability of $\omega \mapsto (\varphi(t, \omega)\mu_\omega)(f)$ follows from Proposition 3.3 (i). Putting both arguments together we get that $\omega \mapsto (\varphi(t, \cdot)\mu_\cdot) \circ \vartheta_{-t}(\omega) = \varphi(t, \vartheta_{-t}\omega)\mu_{\vartheta_{-t}\omega}$ defines a random measure. Denote the measure on $X \times \Omega$ associated with this random measure along Proposition 3.3 (ii) by ν, say. Then, for $B \in \mathscr{B}$ and $F \in \mathscr{F}$,

$$
\begin{aligned}
(\Theta_t\mu)(B \times F) &= \mu\big(\Theta_t^{-1}(B \times F)\big) \\
&= \mu\{(x, \omega) : x \in \varphi(t, \omega)^{-1}B, \ \vartheta_t\omega \in F\} \\
&= \int_\Omega 1_F(\vartheta_t\omega)(\varphi(t, \omega)\mu_\omega)(B) \, dP(\omega) \\
&= \int_\Omega 1_F(\omega)(\varphi(t, \vartheta_{-t}\omega)\mu_{\vartheta_{-t}\omega})(B) \, dP(\omega) \\
&= \nu(B \times F),
\end{aligned}
$$

using invariance of P with respect to ϑ_t for the last but one identity. This holds for every $B \in \mathscr{B}$ and $F \in F$, hence $\Theta_t\mu = \nu$ by Proposition 3.5. Now Proposition 3.6 implies that the disintegrations of $\Theta_t\mu$ and ν coincide for P-almost all ω, which gives (6.5).

In order to verify (6.6), first note that for any $\nu \in Pr_{\Omega,\mathscr{F}_{\leq 0}}(X)$ we have, for $t \geq 0$,

$$\omega \mapsto \varphi(t, \vartheta_{-t}\omega)\nu_\omega \in Pr_{\Omega,\mathscr{F}_{\leq 0}}(X). \tag{6.7}$$

In fact, for any $f \in C(X)$ the bounded real valued map given by $(x, \omega) \mapsto f(\varphi(t, \vartheta_{-t}\omega)x)$ is measurable with respect to $\mathscr{B} \otimes \mathscr{F}_{\leq 0}$. Therefore, $\omega \mapsto \nu_\omega(\varphi(t, \vartheta_{-t}\omega)f) = (\varphi(t, \vartheta_{-t}\omega)\nu_\omega)(f)$ is $\mathscr{F}_{\leq 0}$-measurable, invoking Proposition 3.3 (i) again. From Remark 3.20 (i), applied for $Pr_{\Omega,\mathscr{F}_{\leq 0}}(X)$, we conclude that (6.7) holds true. Thus for $\mu \in Pr_{\Omega,\mathscr{F}_{\leq 0}}(X)$ we get $(\omega \mapsto \mu_{\vartheta_{-t}\omega}) \in Pr_{\Omega,\mathscr{F}_{\leq 0}}(X)$ by (6.4), whence $\omega \mapsto \varphi(t, \vartheta_{-t}\omega)\mu_{\vartheta_{-t}\omega} \in Pr_{\Omega,\mathscr{F}_{\leq 0}}(X)$ from (6.7), thus establishing (6.6). $\quad\square$

As a consequence we get a result which sharpens Corollary 6.13 considerably.

6.17 Theorem (EXISTENCE OF AN INVARIANT MARKOV MEASURE SUPPORTED BY AN INVARIANT COMPACT RANDOM SET) *Suppose that φ is an RDS on a Polish space with two-sided base time. If $\omega \mapsto K(\omega)$ is a forward invariant compact random set which is measurable with respect to the past $\mathscr{F}_{\leq 0}$ of the system, then there exists an invariant Markov measure supported by K. In particular, for any sequence of $\mathscr{F}_{\leq 0}$-measurable random measures σ^n supported by K, and for any increasing sequence of nonnegative random times $t_n : \Omega \to T$, $n \in \mathbb{N}$, which are $\mathscr{F}_{\leq 0}$-measurable and which satisfy $\limsup E(1/t_n) = 0$ for $n \to \infty$, the sequence*

$$\omega \mapsto \gamma_\omega^n = \frac{1}{t_n(\omega)} \int_0^{t_n(\omega)} \Theta_s \sigma_\omega^n \, ds$$

has a convergent subsequence, and every convergent subsequence of $(\gamma^n)_n$ converges to an invariant Markov measure for φ, which is, in addition, supported by K.

PROOF Putting

$$\Gamma = \left\{\mu \in Pr_\Omega(X) : \mu_\omega(K(\omega)) = 1 \; P\text{-a.s.}\right\} \cap Pr_{\Omega,\mathscr{F}_{\leq 0}}(X),$$

and noting that Γ is nonvoid by applying Theorem 2.6 with respect to $(\Omega, \mathscr{F}_{\leq 0})$, the proof proceeds exactly as that of Corollary 6.13 by observing that $Pr_{\Omega,\mathscr{F}_{\leq 0}}(X)$ is closed, invoking Proposition 4.21. $\quad\square$

6.18 Remark Not every invariant measure supported by an $\mathscr{F}_{\leq 0}$-measurable random set K must be a Markov measure. On the contrary, existence of invariant measures which are not Markov is a common property of RDS on compact state spaces (see Crauel [10], 8.3.2, p. 280).

Ergodicity and Extremality of Invariant Measures

We are now going to consider questions concerning ergodicity of invariant measures. Invariant measures for RDS are invariant measures for the skew product flow (see Definition 6.9 (i)). Since the skew product flow has fixed marginal P on Ω, the relations between ergodicity and extremality in the set of invariant measures for an RDS cannot be inferred directly from the corresponding results for measurable dynamical systems.

The following lemma needs only the measurable structure of an RDS φ.

6.19 Lemma *Suppose that φ is an RDS on a Polish space X.*

(i) *The set of invariant measures for φ is convex, and any ergodic invariant measure for φ is an extremal point of the convex set*

$$\left\{\mu \in Pr_\Omega(X) : \mu \text{ invariant for } \varphi\right\}.$$

If P is ergodic for (ϑ_t), then any extremal point of the set $\{\mu \in Pr_\Omega(X) : \mu \text{ invariant for } \varphi\}$ is ergodic.

(ii) *Suppose that $\omega \mapsto C(\omega)$ is a closed random set. If P is ergodic, then any extremal point of*

$$\Gamma = \left\{\mu \in Pr_\Omega(X) : \mu(C) = 1, \ \mu \text{ invariant for } \varphi\right\}$$

is an ergodic invariant measure for φ.

PROOF (i) Convexity of the set of invariant measures for φ is immediate.

Suppose that μ is ergodic. Then μ is an extremal point of the set of all (Θ_t)-invariant measures on $X \times \Omega$ (not necessarily with marginal P on Ω). Consequently, it is also an extremal point of the (convex) subset of (Θ_t)-invariant measures on $X \times \Omega$ with marginal P on Ω.

Next we want to prove that extremal points of $H = \{\mu \in Pr_P(X \times \Omega) : \mu \ (\Theta_t)\text{-invariant}\}$ are (Θ_t)-ergodic (recall that we identify H with the set

$\{\mu \in Pr_\Omega(X) : \mu$ invariant for $\varphi\}$). To do so it suffices to prove that any element of H which is extremal with respect to H is also an extremal point of $\{\nu \in Pr(X \times \Omega) : \nu \ (\Theta_t)$-invariant$\}$. Note that $\Theta_t \nu = \nu$ implies $\vartheta_t(\pi_\Omega \nu) = \pi_\Omega \nu$, which means that any (Θ_t)-invariant measure must have a (ϑ_t)-invariant marginal on Ω. Now suppose that $\mu \in H$ is extremal in H, and suppose that

$$\mu = \alpha \nu_1 + (1 - \alpha)\nu_2$$

for some $0 < \alpha < 1$ and (Θ_t)-invariant $\nu_1, \nu_2 \in Pr(X \times \Omega)$. Then

$$P = \pi_\Omega \mu = \alpha(\pi_\Omega \nu_1) + (1 - \alpha)(\pi_\Omega \nu_2),$$

hence $P = \alpha Q_1 + (1 - \alpha)Q_2$ with (ϑ_t)-invariant $Q_1 = \pi_\Omega \nu_1$ and $Q_2 = \pi_\Omega \nu_2$. But P being ergodic, hence extremal, implies $Q_1 = Q_2 = P$, and thus ν_1 and ν_2 both have marginal P on Ω. This implies $\nu_1, \nu_2 \in H$, so μ being extremal in H entails $\nu_1 = \nu_2 = \mu$, and hence μ is extremal in $\{\mu \in Pr(X \times \Omega) : \mu \ (\Theta_t)$-invariant$\}$. Consequently, μ is ergodic for (Θ_t).

(ii)   Suppose that $\mu \in \Gamma$ is an extremal point with respect to Γ. We want to prove that μ is extremal in the bigger space $H = \{\mu \in Pr_\Omega(X) : \mu$ invariant for $\varphi\}$. This implies that μ is ergodic by (i). If $\mu = \alpha \nu^1 + (1-\alpha)\nu^2$ for some $\nu^1, \nu^2 \in H, 0 < \alpha < 1$, then $\mu(C^c) = 0$ implies $\nu^1(C^c) = \nu^2(C^c) = 0$, hence $\nu^1, \nu^2 \in \Gamma$. But μ was extremal with respect to Γ, so $\nu^1 = \nu^2 = \mu$. $\square$

We note that ergodicity of the base system (ϑ_t, P) is a necessary condition for the existence of an ergodic invariant measure for φ.

We will need an elementary lemma about extremal points of convex sets. Roughly it says that every extremal point of the image of a compact convex set under a continuous convex map is the image of an extremal point of the original set. Imagine the unit ball in the L^1-norm of $\mathbb{R}^2$ with the map being projection to either of the coordinate axes to see that not every image of an extremal point of the original set must be an extremal point of the image. Imagine the unit ball in the L^∞-norm of $\mathbb{R}^2$ with the same map to see that also non-extremal points of the original set may be mapped to extremal points of the image.

6.20 Lemma *Suppose that K is a compact convex set with a locally convex topology, and denote by $\mathcal{E}(K)$ the set of its extremal points. Let $a : K \to B$ be a continuous convex map from K to a locally convex topological vector space B. Then*

$$\mathcal{E}(a(K)) \subset a(\mathcal{E}(K)),$$

where $\mathcal{E}(a(K))$ denotes the set of extremal points of $a(K)$, the (compact convex) image of K under a.

PROOF   Pick an extremal point $e \in \mathcal{E}(a(K))$ (a convex compact subset of a locally convex vector space has extremal points; cf. Dunford and Schwartz [20], Lemma V.8.2, p. 439). We want to prove that there exists an extremal point of K which is mapped to e under a. The set $a^{-1}\{e\}$ is nonvoid, convex, and closed, and hence compact. Let u be an extremal point of $a^{-1}\{e\}$ (with respect to $a^{-1}\{e\}$; extremal points of $a^{-1}\{e\}$ exist by virtue of the result mentioned above). We claim that u is an extremal point of K, and, of course, $a(u) = e$. In fact, if $u = \alpha v_1 + (1 - \alpha)v_2$ for some $v_1, v_2 \in K$, $0 < \alpha < 1$, then $a(u) = \alpha a(v_1) + (1 - \alpha)a(v_2) = e$. Since e was assumed to be an extremal point of $a(K)$, this implies $a(v_1) = a(v_2) = e$, and hence $v_1, v_2 \in a^{-1}\{e\}$. Now u is an extremal point of $a^{-1}\{e\}$, so $u = \alpha v_1 + (1 - \alpha)v_2$ implies $v_1 = v_2 = u$. This proves that u is an extremal point of K. $\square$

Oxtoby Theory for Random Dynamical Systems

For a continuous transformation of a compact metrisable space the following assertions are equivalent:

- There exists only one invariant measure (which is then necessarily ergodic); in this case the system is said to be *uniquely ergodic*.

- For every continuous function the time mean of the function evaluated along the orbit through an arbitrary initial point converges to a constant, which is independent of the initial point.

- For every continuous function the sequence of continuous functions, given by the time means of the function evaluated along the orbit, converges to a constant function in the topology of uniform convergence on the space of continuous functions.

This is the theorem of Oxtoby [31], which we cited here following Walters [38], Theorem 6.19, p. 160. This result can be generalised in several respects.

Suppose that $\Gamma \subset Pr_\Omega(X)$ is compact and connected, and suppose that $f \in C_\Omega(X)$. Since $\mu \mapsto \mu(f)$ is continuous, the set $\{\mu(f) : \mu \in \Gamma\} \subset \mathbb{R}$ is the compact interval

$$\left[\min\{\mu(f) : \mu \in \Gamma\}, \max\{\mu(f) : \mu \in \Gamma\}\right]. \tag{6.8}$$

Next suppose that Γ is not only connected but also convex. The end points of the interval being its extremal points, Lemma 6.20 then implies that both end points of (6.8) are taken by extremal elements of Γ.

Now suppose that φ is an RDS over an ergodic base system $((\vartheta_t), P)$, suppose further that $\omega \mapsto K(\omega)$ is a φ-invariant compact random set, and consider

$$\Gamma = \big\{ \mu \in Pr_\Omega(X) : \mu(K) = 1, \ \mu \text{ invariant for } \varphi \big\}.$$

Then Γ is compact and convex, and its extremal points are ergodic by Lemma 6.19 (ii). Since the end points of the interval in (6.8) are realised by extremal points of Γ we thus have proved that, for every $f \in C_\Omega(X)$, there exist ergodic invariant measures $\mu^{\min} \in \Gamma$ and $\mu^{\max} \in \Gamma$ such that

$$\int_{X \times \Omega} f \, d\mu^{\min} = \min\Big\{ \int_{X \times \Omega} f \, d\mu : \mu(K) = 1, \ \mu \text{ invariant for } \varphi \Big\},$$

$$\int_{X \times \Omega} f \, d\mu^{\max} = \max\Big\{ \int_{X \times \Omega} f \, d\mu : \mu(K) = 1, \ \mu \text{ invariant for } \varphi \Big\}.$$

Note that $\mu^{\min}$ and $\mu^{\max}$ depend on f, and that they need not be unique.

6.21 Proposition (THE SUPREMUM OF TIME MEANS IS REALISED BY AN ERGODIC INVARIANT MEASURE) *Suppose that φ is an RDS over an ergodic base system (ϑ_t, P), and that $\omega \mapsto K(\omega)$ is a φ-invariant compact set. Let $f : X \times \Omega \to \mathbb{R}$ be measurable, with $x \mapsto f(x, \omega)$ continuous on $K(\omega)$ P-a.s., and*

$$\omega \mapsto \sup_{x \in K(\omega)} f^+(x, \omega) \tag{6.9}$$

integrable with respect to P. Then the following limit exists P-a.s., and satisfies

$$\lim_{t \to \infty} \frac{1}{t} \Big(\sup_{x \in K(\omega)} \int_0^t f \circ \Theta_s(x, \omega) \, ds \Big)$$
$$= \max\Big\{ \int_{X \times \Omega} f \, d\mu : \mu(K) = 1, \ \mu \text{ invariant for } \varphi \Big\}. \tag{6.10}$$

In particular, there exists an ergodic φ-invariant measure μ, depending on f, such that

$$\lim_{t \to \infty} \frac{1}{t} \Big(\sup_{x \in K(\omega)} \int_0^t f \circ \Theta_s(x, \omega) \, ds \Big) = \int_{X \times \Omega} f(x, \omega) \, d\mu(x, \omega)$$

P-a.s. (the integral is to be understood as a sum in case time T is discrete).

PROOF Put

$$F_t(\omega) = \left(\sup_{x \in K(\omega)} \int_0^t f \circ \Theta_s(x, \omega) \, ds \right)$$

and $F(\omega) = \limsup_{t \to \infty} \frac{1}{t} F_t(\omega)$. From (6.9) we get $F(\omega) < \infty$ P-a.s., since

$$\sup_{x \in K(\omega)} \int_0^t (f \circ \Theta_s)(x, \omega) \, ds \leq \int_0^t \sup_{x \in K(\omega)} f(x, \cdot) \circ \vartheta_s(\omega) \, ds.$$

A straightforward computation, using that $\varphi(\tau, \omega)K(\omega) \subset K(\vartheta_\tau \omega)$ P-a.s. for every $\tau \geq 0$, yields $F_{t+\tau}(\omega) \leq F_\tau(\omega) + F_t(\vartheta_\tau \omega)$ P-a.s., whence $F(\omega) \leq F(\vartheta_\tau \omega)$ P-a.s., using $F(\omega) < \infty$ P-a.s. This implies $F(\omega) = F(\vartheta_\tau \omega)$ (see Proposition B.6 (i)). Therefore $F(\omega)$ is a constant P-a.s., which we denote by F again. We note that one even gets $F = \lim \frac{1}{t} F_t(\omega) = \inf \frac{1}{t} F_t(\omega)$ P-a.s. as a consequence of the subadditive ergodic theorem of Kingman [28], but we will not make use of this here. We have either $F = -\infty$, or $F > -\infty$. Since for every ergodic φ-invariant μ with $\mu(K) = 1$ we get

$$\int_{X \times \Omega} f(y, \omega) \, d\mu(y, \omega) = \lim_{t \to \infty} \frac{1}{t} \int_0^t f \circ \Theta_s(y, \omega) \, ds \leq F$$

for μ-almost all $(y, \omega) \in X \times \Omega$ from the individual ergodic theorem, there is nothing to prove in case $F = -\infty$. Suppose that $F > -\infty$. For $\varepsilon > 0$ and $n \in \mathbb{N}$ consider the set

$$\left\{ (t, x, \omega) \in T \times X \times \Omega : t \geq n \text{ and } \frac{1}{t} \int_0^t (f \circ \Theta_s)(x, \omega) \, ds \geq F - \varepsilon \right\}.$$

This set is measurable in $T \times X \times \Omega$, and since

$$\left\{ (t, x) : t \geq n \text{ and } \frac{1}{t} \int_0^t (f \circ \Theta_s)(x, \omega) \, ds \geq F - \varepsilon \right\}$$

is a nonvoid closed set P-a.s., we can choose, for every $n \in \mathbb{N}$, a measurable selection $\omega \mapsto (t_n(\omega), x_n(\omega))$. Then

$$\frac{1}{t_n(\omega)} \int_0^{t_n(\omega)} f \circ \Theta_s(x_n(\omega), \omega) \, ds \geq F - \varepsilon,$$

and $t_n(\omega) \geq n$, both P-a.s. Define measures $\sigma_\omega^n = \delta_{x_n(\omega)}$ and

$$\gamma_\omega^n = \frac{1}{t_n(\omega)} \int_0^{t_n(\omega)} (\Theta_s \sigma^n)_\omega \, ds$$

(both $(\sigma^n)_{n\in\mathbb{N}}$ and $(\gamma^n)_{n\in\mathbb{N}}$ depend on ε). By Corollary 6.13 there exists a subsequence $(\gamma^{n_k})_{k\in\mathbb{N}}$ converging to an invariant measure γ^ε for φ, and $\gamma^\varepsilon(K) = 1$. Upper semi-continuity of $\gamma \mapsto \int f \, d\gamma$ together with the definition of γ^n gives

$$
\int f \, d\gamma^\varepsilon \;\geq\; \limsup_{k\to\infty} \int f \, d\gamma^{n_k}
$$
$$
= \; \limsup_{k\to\infty} E\left(\frac{1}{t_{n_k}} \int_0^{t_{n_k}} f \circ \Theta_s(x_n(\cdot), \cdot) \, ds \right)
$$
$$
\geq \; F - \varepsilon. \tag{6.11}
$$

Now put $R_f = \sup\{ \int f \, d\gamma : \gamma(K) = 1, \; \gamma \text{ invariant for } \varphi \}$, then $R_f \geq F - 1$, say. We know that $\{ \gamma : \gamma(K) = 1, \; \gamma \; \varphi\text{-invariant} \}$ is compact (by Proposition 4.3 together with Lemma 6.7) and nonvoid (by Corollary 6.13). By Lemma 3.19, $\mu \mapsto \int f \, d\mu$ is upper semi-continuous on $\{ \mu : \mu(K) = 1 \}$, whence $\{ \mu : \mu(K) = 1, \; \mu \; \varphi\text{-invariant, and } \int f \, d\mu \geq F - 1 \}$ is compact and nonvoid. Thus R_f is realised as a maximum instead of just a supremum, which implies, in particular,

$$
\left\{ \mu : \mu \; \varphi\text{-invariant}, \; \mu(K) = 1, \; \int f \, d\mu = R_f \right\} \neq \emptyset.
$$

Since extremal points of this compact convex set are extremal in the set $\{ \nu : \nu \; \varphi\text{-invariant} \}$, whence ergodic, there exists an ergodic φ-invariant μ supported by K with $\int f \, d\mu = R_f$. Finally, invoking (6.11),

$$
\int f \, d\mu = R_f \geq \int f \, d\gamma^\varepsilon \geq F - \varepsilon
$$

for $\varepsilon > 0$ arbitrary gives $\int f \, d\mu = R_f = F$, which is the assertion. $\qquad\square$

Since random continuous functions satisfy the condition of Proposition 6.21, we obtain that for every $f \in C_\Omega(X)$, and for every φ-invariant compact set $\omega \mapsto K(\omega)$, there exists an ergodic invariant measure μ, supported by K, such that (6.10) is satisfied. But it should be kept in mind that in general μ will depend on the function f under consideration.

6.22 Corollary *Suppose that, under the assumptions of Proposition 6.21, $f \in C_\Omega(X)$ has the property that*

$$
\mu \mapsto \int f \, d\mu
$$

*is independent of μ, and hence equal to a constant $r(f)$, say.   Then for
P-almost every ω,*

$$\frac{1}{t}\int_0^t f \circ \Theta_s(x,\omega)\,ds \longrightarrow r(f)$$

for $t \to \infty$, uniformly in $x \in K(\omega)$.
*This holds, in particular, if φ is uniquely ergodic on K, i.e., if there is only
one invariant measure for φ supported by K.*

6.23 Remark 'Uniform convergence for P-almost every ω' means that there
is a set of full P-measure, such that for ω from this set the convergence is
uniform in x. It does not imply that the convergence is uniform in ω.

6.24 Corollary (OXTOBY THEOREM FOR RDS) *If φ is an RDS over
an ergodic base system (ϑ_t, P) on a compact metrisable space X, then the
following three assertions are equivalent:*

(i) *The RDS φ is uniquely ergodic.*

(ii) *For every $f \in C_\Omega(X)$ the time means*

$$\frac{1}{t}\int_0^t f \circ \Theta_s(x,\omega)\,ds$$

*converge P-a.s., for $t \to \infty$, to a random variable $r(f,\omega)$, say, inde-
pendently of $x \in X$.*

(iii) *For every $f \in C_\Omega(X)$ the family F_t of random continuous functions,
given by*

$$F_t(x,\omega) = \frac{1}{t}\int_0^t f \circ \Theta_s(x,\omega)\,ds$$

*converges P-a.s. to a constant $r(f)$, say, in the supremum norm on
bounded continuous functions on X.*

6.25 Remark Clearly the assertions of Proposition 6.21 and of its corollaries
do not hold true if the function f under consideration is not continuous in
$x \in X$, but only measurable.

Appendix A

The Narrow Topology on Non-Random Measures

We reprove the (well known) fact that the narrow topology on the space $Pr(X)$ of Borel probability measures on a Polish space X is Polish itself, and in particular metrisable. This result is due to Prohorov [33]. A proof can be found in several places. We follow Dudley [19], Section 11.3, closely. In fact, Theorem A.2 is the same as Theorem 11.3.3 of Dudley [19], except that here the result is formulated as coincidence of the four topologies. Dudley formulates only coincidence of convergence of sequences. This is sufficient as soon as one knows that all the four topologies are metrisable (which they are). However, equality of two topologies does not follow from coincidence of all convergent sequences as soon as one of the topologies is not metrisable (see Example A.3). The proof of Theorem A.2 essentially remains the same as that of Theorem 11.3.3 of Dudley.

Let X be a Polish space equipped with a complete metric d, and $\mathscr{B}$ denoting the Borel sets. For an arbitrary $A \subset X$ and $\delta > 0$ denote by A^δ the δ-neighbourhood of A,

$$A^\delta = \{x \in X : d(x, A) < \delta\} = \{x : d(x, a) < \delta \text{ for some } a \in A\}.$$

Recall that the narrow topology of $Pr(X)$ is the smallest topology such that $\rho \mapsto \int f \, d\rho$ is continuous for every bounded continuous $f : X \to \mathbb{R}$. Denote the space of bounded continuous functions from X to $\mathbb{R}$ by $C_b(X)$. For the purposes of this appendix the narrow topology is denoted by $\mathscr{T}_1$. A neighbourhood basis at $\rho \in Pr(X)$ for $\mathscr{T}_1$ is given by the collection of all sets

$$V_{f_1,\ldots,f_n;\delta}(\rho) = \left\{ \zeta \in Pr(X) : |\int f_k \, d\rho - \int f_k \, d\zeta| < \delta, k = 1, \ldots, n \right\} \quad \text{(A.1)}$$

with $n \in \mathbb{N}$, $f_1, \ldots, f_n \in C_b(X)$, and $\delta > 0$.

We will make use of the following lemma.

A.1 Lemma *The narrow topology is generated by $\{f \in C_b(X) : 0 \le f \le 1\}$ already.*

PROOF   We prove that continuity of $\rho \mapsto \int f \, d\rho$ for every $f \in C_b(X)$ with $0 \le f \le 1$ implies continuity of $\rho \mapsto \int h \, d\rho$ for $h \in C_b(X)$ arbitrary. In fact, for h constant the assertion is immediate. If h is not constant, put $f(x) = (\sup h - \inf h)^{-1}(h(x) - \inf h)$, where $\sup h = \sup_{x \in X} h(x)$, $\inf h = \inf_{x \in X} h(x)$. Then $0 \le f \le 1$, and

$$\int h \, d\rho = (\sup h - \inf h) \int f \, d\rho + \inf h.$$

With $\rho \mapsto \int f \, d\rho$ continuous, also $\rho \mapsto \int h \, d\rho$ is continuous.   □

A function $g \in C_b(X)$ is *Lipschitz* if

$$\|g\|_L = \sup_{x \ne y} \frac{|g(x) - g(y)|}{d(x, y)} < \infty.$$

With $\|g\|_{\mathrm{BL}} = \|g\|_L + \sup_{x \in X} |g(x)|$ the space

$$\mathrm{BL}(X) = \mathrm{BL}(X, d) = \{g : X \to \mathbb{R} : \|g\|_{\mathrm{BL}} < \infty\}$$

is a Banach space. Elements are addressed to as bounded Lipschitz functions on X.

Introduce another topology, denoted by $\mathscr{T}_2$, to be the smallest topology such that $\rho \mapsto \int g \, d\rho$ is continuous for every $g \in \mathrm{BL}(X)$. A neighbourhood basis at $\rho \in Pr(X)$ for $\mathscr{T}_2$ is the collection of sets $V_{g_1, \ldots, g_n; \delta}(\rho)$ (as defined in (A.1)) with $g_1, \ldots, g_n \in \mathrm{BL}(X)$ for some $n \in \mathbb{N}$, and $\delta > 0$. The definition of $\mathscr{T}_2$ refers to the concrete metric d on X, whereas that of $\mathscr{T}_1$ does not. Clearly $\mathscr{T}_1$ is finer than $\mathscr{T}_2$.

Define metrics α and β on $Pr(X)$ by

$$\alpha(\rho, \zeta) = \inf\{\eta > 0 : \rho(B) \le \zeta(B^\eta) + \eta \text{ for all } B \in \mathscr{B}\}, \qquad (\mathrm{A}.2)$$

where $B^\eta = \{x \in X : d(x, B) < \eta\}$ denotes the η-neighbourhood of B, and

$$\beta(\rho, \zeta) = \sup\{|\textstyle\int g \, d\rho - \int g \, d\zeta| : \|g\|_{\mathrm{BL}} \le 1\}. \qquad (\mathrm{A}.3)$$

It is immediate that β defines a metric. Also α defines a metric, referred to as the *Prohorov metric*, but this fact is not that immediate (see Dudley [19], Theorem 11.3.1, p. 309). Denote by $\mathscr{T}_3$ the metric topology induced by β, and by $\mathscr{T}_4$ the metric topology generated by α. A basis for a metric topology is given by the collection of all balls. Note that also $\mathscr{T}_3$ and $\mathscr{T}_4$ refer to the choice of the metric d on X. Both the metrics α and β are complete (see Corollary 11.5.5 of Dudley [19], p. 317).

Note that, for ρ and ζ in $Pr(X)$ and $B \in \mathscr{B}$ arbitrary,

$$\alpha(\rho, \zeta) < \eta \text{ implies } \zeta(B) \leq \rho(B^\eta) + \eta. \tag{A.4}$$

Recall that a topology $\mathscr{T}$ is finer than another topology $\mathscr{T}'$ if for every element V' of a basis of $\mathscr{T}'$ there exists an element $U \in \mathscr{T}$ with $U \subset V'$.

A.2 Theorem $\mathscr{T}_1 = \mathscr{T}_2 = \mathscr{T}_3 = \mathscr{T}_4$, where

(i) $\mathscr{T}_1$ is the narrow topology,

(ii) $\mathscr{T}_2$ is the smallest topology such that $\rho \mapsto \int g \, d\rho$ is continuous for every $g \in \mathrm{BL}(X)$,

(iii) $\mathscr{T}_3$ is the topology induced by the metric α, defined in (A.3),

(iv) $\mathscr{T}_4$ is the topology induced by the Prohorov metric α, defined in (A.2).

PROOF Clearly $\mathscr{T}_1$ is finer than $\mathscr{T}_2$.
To prove that $\mathscr{T}_2$ is finer than $\mathscr{T}_3$ fix $\rho \in Pr(X)$. The assertion follows if there are $g_0, g_1, \ldots, g_n \in \mathrm{BL}(X)$, $n \in \mathbb{N}$, such that $| \int g_k \, d\rho - \int g_k \, d\zeta | < \varepsilon/11$ for $k = 1, \ldots, n$ implies $\zeta \in B_\varepsilon^\beta(\rho)$, where $B_\varepsilon^\beta(\rho)$ is the ball of radius $\varepsilon > 0$ around $\rho \in Pr(X)$ with respect to β,

$$B_\varepsilon^\beta(\rho) = \{\zeta \in Pr(X) : \beta(\rho, \zeta) < \varepsilon\}.$$

This implies $V_{g_0, g_1, \ldots, g_n; \varepsilon/11}(\rho) \subset B_\varepsilon^\beta(\rho)$, establishing the fact that $\mathscr{T}_2$ is finer that $\mathscr{T}_3$. Put $\delta = \varepsilon/11$ (for brevity of notation). Let $K \subset X$ be a compact set with $\rho(K) \geq 1 - \delta$. Consider the set $B = \{h \in \mathrm{BL}(X) : \|h\|_{\mathrm{BL}} \leq 1\}$. Then $\{h|_K : h \in B\}$ is a compact subset of $C(K)$ with respect to the supremum topology. So there exist $g_1, \ldots, g_n \in B$ such that for every $h \in B$ there is a j with

$$\sup_{z \in K} |h(z) - g_j(\dot{z})| < \delta \text{ (and, of course, } \sup_{x \in X} |h(x) - g_j(x)| \leq 2). \tag{A.5}$$

We note that (A.5) implies

$$\sup_{z \in K^\delta} |h(z) - g_j(z)| < 3\delta, \tag{A.6}$$

since for $z \in K$ and $d(x, z) < \delta$

$$
\begin{aligned}
|h(x) - g_j(x)| &\leq |h(x) - h(z)| + |h(z) - g_j(z)| + |g_j(z) - g_j(x)| \\
&\leq \|h\|_L d(x, z) + \delta + \|g_j\|_L d(x, z) < 3\delta.
\end{aligned}
$$

Finally, let g_0 be a bounded Lipschitz function with $1_K \leq g_0 \leq 1_{K^\delta}$. For example, put $g_0(x) = \max\{0, 1 - d(x, K)/\delta\}$. Then, for every $\zeta \in Pr(X)$ with $\int g_0 \, d\rho - \int g_0 \, d\zeta < \delta$, we have

$$
\zeta(K^\delta) \geq \int g_0 \, d\zeta > \int g_0 \, d\rho - \delta \geq \rho(K) - \delta \geq 1 - 2\delta,
$$

hence $\zeta(K^{\delta c}) \leq 2\delta$, where $K^{\delta c} = (K^\delta)^c$.

We now claim that any $\zeta \in Pr(X)$ with $|\int g_k \, d\rho - \int g_k \, d\zeta| < \delta$ for $0 \leq k \leq n$ satisfies $\beta(\rho, \zeta) < \varepsilon$. In fact, pick $h \in B$, and let j, $0 \leq j \leq n$, be such that (A.5) (and consequently (A.6)) are satisfied with h and g_j. Then

$$
\begin{aligned}
|\textstyle\int h \, d\rho - \int h \, d\zeta| &\leq |\textstyle\int (h - g_j) \, d\rho| + |\textstyle\int (h - g_j) \, d\zeta| + |\textstyle\int g_j \, d\rho - \int g_j \, d\zeta| \\
&< |\textstyle\int (h - g_j) \, d\rho| + |\textstyle\int (h - g_j) \, d\zeta| + \delta \\
&\leq |\textstyle\int_K (h - g_j) \, d\rho| + |\textstyle\int_{K^c} (h - g_j) \, d\rho| \\
&\quad + |\textstyle\int_{K^\delta} (h - g_j) \, d\zeta| + |\textstyle\int_{(K^\delta)^c} (h - g_j) \, d\zeta| + \delta \\
&\leq \delta + 2\rho(K^c) + 3\delta + 2\zeta(K^{\delta c}) + \delta \\
&\leq \delta + 2\delta + 3\delta + 4\delta + \delta = 11\delta.
\end{aligned}
$$

Consequently, the open $\mathcal{T}_2$-neighbourhood $V_{g_0, g_1, \ldots, g_n; \delta}(\rho)$ with $\delta = \varepsilon/11$ is contained in $B_\varepsilon^\beta(\rho)$. Hence $\mathcal{T}_2$ is finer than $\mathcal{T}_3$.

Next, $\mathcal{T}_3$ is finer than $\mathcal{T}_4$, since $\alpha^2 \leq 2\beta$ (see Dudley [19], Problem 5 (b) in Section 11.3, p. 312).

Finally, to prove that $\mathcal{T}_4$ is finer that $\mathcal{T}_1$ it suffices to show that for every $f \in C_b(X)$, every $\varepsilon > 0$, and every $\rho \in Pr(X)$ there exists $\delta > 0$ such that the δ-ball around ρ in the α-metric is contained in $V_{f; \varepsilon}(\rho)$, i.e., $|\int f \, d\rho - \int f \, d\zeta| < \varepsilon$ for every ζ with $\alpha(\rho, \zeta) < \delta$.

By Lemma A.1 it suffices to prove this for $f \in C_b(X)$ with $0 \leq f \leq 1$. Choose $n \in \mathbb{N}$ with $n > 3/\varepsilon$. Put $F_k = \{x \in X : f(x) \geq k/n\}$. Then F_k is closed, and hence there exists $\delta_0 > 0$ such that $\rho(F_k^\delta) \leq \rho(F_k) + \varepsilon/6$ for $0 \leq k \leq n$ and every $\delta \leq \delta_0$. From Lemma 1.4 we infer for any ζ with $\alpha(\rho, \zeta) < \delta_1 = \min\{\delta_0, \varepsilon/6\}$, invoking (A.4),

$$\int f \, d\zeta \;\le\; \frac{1}{n}\sum_{k=0}^{n}\zeta(F_k) \;\le\; \frac{1}{n}\sum_{k=0}^{n}\left(\rho(F_k^\delta)+\delta\right)$$

$$\le\; \frac{1}{n}\sum_{k=0}^{n}\left(\rho(F_k)+\varepsilon/6\right)+\frac{n+1}{n}\delta$$

$$=\; \frac{1}{n}\left(1+\sum_{k=1}^{n}\rho(F_k)\right)+\frac{n+1}{n}(\varepsilon/6+\delta)$$

$$\le\; \frac{1}{n}\sum_{k=1}^{n}\rho(F_k)+\frac{1}{n}+\frac{n+1}{n}(\varepsilon/6+\delta)$$

$$<\; \int f \, d\rho+\varepsilon/3+2(\varepsilon/6+\varepsilon/6)$$

$$=\; \int f \, d\rho+\varepsilon.$$

Applying the above argument to $1-f$ instead of f yields

$$\int (1-f)\, d\zeta < \int (1-f)\, d\rho+\varepsilon,$$

hence

$$\int f \, d\rho < \int f \, d\zeta + \varepsilon$$

for all ζ with $\alpha(\rho,\zeta)\le\delta_2$. This implies the assertion for $\delta=\min\{\delta_1,\delta_2\}$. $\square$

A.3 Example   This is an example of two distinct topologies such that convergence of a sequence in one of them is equivalent to convergence of the sequence in the other one.

Consider $\mathbb{N}^2 = \mathbb{N}\times\mathbb{N}$, equipped with the following topology:

- $\mathbb{N}\times\mathbb{N}\setminus\{(0,0)\}$ has the discrete topology

- $U\subset\mathbb{N}\times\mathbb{N}$ is a neighbourhood of $(0,0)$ if $(0,0)\in U$ and if the sets $\{n\in\mathbb{N}:(n,m)\notin U\}$ are finite for almost all $m\in\mathbb{N}$.

Then there exists no sequence in $\mathbb{N}\times\mathbb{N}$ converging to $(0,0)$. In fact, suppose that $(\gamma^k)_{k\in\mathbb{N}}$ is a sequence. We distinguish two cases. Either there exists $m_0\in\mathbb{N}$ such that $\gamma_2^k=m_0$ for infinitely many k, where γ_2^k denotes the second component of γ^k. In this case put $U=\{(n,m):m\neq m_0\}$, which gives an open neighbourhood of $(0,0)$ such that $\gamma^k\notin U$ for infinitely many k, so γ^k cannot converge to $(0,0)$. If, on the other hand, for every $m \in \mathbb{N}$ we have $\gamma_2^k = m$ only for finitely many k, then put $U = \{\gamma^k : k \in \mathbb{N}\}^c$. This gives an open neighbourhood of $(0,0)$ such that $\gamma^k \notin U$ for every $k \in \mathbb{N}$, which also forbids γ^k to converge to $(0,0)$. Consequently, the only convergent sequences with respect to this topology are those which are eventually constant.

Now denote by d the restriction of the standard $\mathbb{R}^2$-metric $(x,y) \mapsto |x-y|$ on $\mathbb{N} \times \mathbb{N}$. The topology induced by d on $\mathbb{N} \times \mathbb{N}$ is the discrete topology. Every convergent sequence with respect to this topology is eventually constant. Consequently, a sequence converges in one of the two topologies if and only if it converges also in the other. But clearly the two topologies do not coincide. This shows that it does not suffice to establish that convergence of sequences is equivalent in two topologies to conclude that the topologies coincide.

The topology used in the above example can be found in Querenburg [34], Aufgabe 5.2, p. 60.

Appendix B

Scattered Results

Here we collect some results which do not fit into the main body, being too remote or too particular.

The following considerations are concerned with a topological result. Recall that for a map τ mapping a topological space X to a set Y the *final topology induced by π on Y* is, by definition, the collection of all sets $U \subset Y$ such that $\tau^{-1}U \subset X$ is open. It is the finest topology on Y such that τ is continuous. If τ is surjective, this topology is also addressed to as the *quotient topology* (since Y can be identified with the quotient X/τ, given by identifying two points x and z in X if $\tau(x) = \tau(z)$).

B.1 Proposition *Suppose that X is a topological space and $\pi : X \to X$ satisfies $\pi^2 = \pi$, where $\pi^2 = \pi \circ \pi$. Put $Y = \pi(X) \subset X$. Then the trace topology on Y is finer than the final topology induced by π on Y.*

PROOF Note that $\pi^2 = \pi$ implies $\pi|_Y = \mathrm{id}$. Thus for every $W \subset Y$ we get $\pi(W) = W$, hence $W \subset \pi^{-1}(W)$, and thus $W \subset \pi^{-1}(W) \cap Y$. On the other hand, for $w \in \pi^{-1}(W) \cap Y$ we have $\pi(w) \in W$ as well as $\pi(w) = w$, and hence $w \in W$. Consequently, every $W \subset Y$ satisfies

$$W = \pi^{-1}(W) \cap Y. \tag{B.1}$$

Now let $U \subset Y$ be open in the final topology induced by π, which means $\pi^{-1}U$ is open in X. Since $U = \pi^{-1}U \cap Y$ by (B.1), U is open with respect to the trace topology of Y. $\qquad \square$

B.2 Corollary *Under the assumptions of the Proposition, π is continuous with respect to the trace topology if and only if the quotient topology on Y coincides with the trace topology on Y.*

PROOF By definition, π is continuous with respect to the quotient topology, so it is continuous with respect to the trace topology in case both coincide.

On the other hand, if π is continuous with respect to the trace topology, then the quotient topology is finer than the trace topology (since the quotient is the finest topology on Y with respect to which π is continuous). By Proposition B.1 the trace topology is finer than the quotient topology, and hence both coincide.   □

B.3 Lemma *Suppose that Y is a topological space, $Z \subset Y$ is an arbitrary subset equipped with the trace topology, and $C \subset Z$. Then C is compact as a subset of Z in its trace topology if and only if C is compact as a subset of Y with the original topology.*

PROOF Denote by $e : Z \to Y$ the embedding. Then e is continuous with respect to the trace topology on Z and the original topology on Y (by the definition of the trace topology). Hence for $C \subset Z$ compact in the trace topology $e(C) = C$ is compact in the original topology. On the other hand, for $C \subset Z$ with $C \subset Y$ compact in the original topology of Y, let

$$C = \bigcup_\alpha (U_\alpha \cap Z)$$

be a covering of C by open sets in the trace topology. Then $C \subset \bigcup_\alpha U_\alpha$, so that compactness of C with respect to the original topology implies existence of a finite subcover $C \subset \bigcup_{1 \leq k \leq n} U_{\alpha_k}$, whence $C = C \cap Z \subset \bigcup_{1 \leq k \leq n} (U_{\alpha_k} \cap Z)$. Thus every cover of C by open sets of the trace topology has a finite subcover, so C is compact in the trace topology.   □

The above argument does not work as soon as one tries to prove that compactness of $C' \subset Y$ in the Y topology implies compactness of $C' \cap Z$ in the trace topology of Z. In fact, this is not true. Every embedding of a non-compact Polish space into a compact metric space gives a counterexample (taking C' to be the ambient compact space). Essentially the same counterexample is given by taking Z to be the space of random probability measures on a non-compact Polish space and Y to be the (compact) space $\mathfrak{C}$ introduced in Theorem 4.14.

The next lemma shows that the measure algebra of a non-atomic probability space is never totally bounded (or compact, which is the same since the measure algebra is complete).

B.4 Lemma *Suppose that* $(\Omega, \mathscr{F}, P)$ *is a non-atomic probability space. Then for every finite collection* $F_1, \ldots, F_n \in \mathscr{F}$ *there exists* $G \in \mathscr{F}$ *such that*

$$P(G \bigtriangleup F_k) = \frac{1}{2}$$

for each k, $1 \le k \le n$.

PROOF For $F_1, \ldots, F_n \in \mathscr{F}$ denote by $\alpha(F_1, \ldots, F_n)$ the algebra generated by $F_1, \ldots, F_n$, and denote by $\{B_1, \ldots, B_N\}$ the partition associated with $\alpha(F_1, \ldots, F_n)$ (characterised by $\alpha(B_1, \ldots, B_N) = \alpha(F_1, \ldots, F_n)$). For each k choose $B'_k \subset B_k$ with $P(B'_k) = P(B_k)/2$; this is possible since $(\Omega, \mathscr{F}, P)$ is non-atomic. Put

$$G = \bigcup_{k=1}^{N} B'_k.$$

For any $A \in \alpha(F_1, \ldots, F_n)$ there exists $J \subset \{1, \ldots, N\}$ such that $A = \bigcup_{j \in J} B_j$, and hence

$$G \bigtriangleup A = \bigcup_{j \in J}(B_j \backslash B'_j) \cup \bigcup_{j \notin J} B'_j.$$

Thus

$$P(G \bigtriangleup A) = \sum_{j \in J} P(B_j \backslash B'_j) + \sum_{j \notin J} P(B'_j) = \frac{1}{2}.$$

This holds for every $A \in \alpha(F_1, \ldots, F_n)$, and hence it holds, in particular, for $F_1, \ldots, F_n$. $\qquad \Box$

The following lemma is concerned with relations between two of the different metrics metrising convergence in probability. These metrics can be found in many textbooks. However, usually only the fact that convergence of sequences in probability is equivalent to convergence in these metrics is established, which, of course, implies that they induce the same topology. Equivalence of the metrics themselves is usually not established, which is what we want to use. Again we restrict ourselves to probability spaces.

B.5 Lemma *Suppose that* $(\Omega, \mathscr{F}, P)$ *is a probability space, and that* $g : \Omega \to [0, \infty]$ *is measurable. Put*

$$\varkappa(g) = \inf\{\alpha \ge 0 : P(g > \alpha) \le \alpha\}$$

and

$$\gamma(g) = \int \min\{1, g\} \, dP.$$

Then

$$\frac{1}{2}\gamma(g) \le \varkappa(g) \le \gamma(g)^{1/2} \le (Eg)^{1/2},$$

where Eg denotes expectation of g.

PROOF Suppose that α satisfies $P(g > \alpha) \le \alpha$. Then

$$\gamma(g) = \int_{\{g > \alpha\}} \min\{1, g\} \, dP + \int_{\{g \le \alpha\}} \min\{1, g\} \, dP \le P(g > \alpha) + \alpha \le 2\alpha.$$

This holds for every α with $P(g > \alpha) \le \alpha$, and hence $\gamma(g) \le 2\inf\{\alpha \ge 0 : P(g > \alpha) \le \alpha\}$, which is the first inequality.

If $\gamma(g) = 0$ then $g = 0$ P-a.s., so $\varkappa(g) = 0$. For any $0 < z \le 1$ the Chebyshev inequality yields

$$P(g > z) \le \frac{1}{z} \int \min\{1, g\} \, dP = \frac{1}{z}\gamma(g),$$

which gives, with $z = \gamma(g)^{1/2}$,

$$P(g > \gamma(g)^{1/2}) \le \gamma(g)^{1/2},$$

whence $\gamma(g)^{1/2} \in \{\alpha : P(g > \alpha) \le \alpha\}$, which gives the second inequality. The third inequality is immediate.   □

Finally, we collect some technical results on integrability and time means. For a real valued function f we write $f = f^+ - f^-$, where $f^+ = f \vee 0 = \max\{f, 0\}$, and $f^- = (-f) \vee 0 = -\min\{f, 0\}$.

B.6 Proposition *Suppose that θ is a measure preserving transformation of a probability space $(\Omega, \mathscr{F}, P)$.*

   (i) *Let $f : \Omega \to \bar{\mathbb{R}} \; (= \mathbb{R} \cup \{-\infty, \infty\})$ be measurable. Then $f \circ \theta \le f$ (P-a.s.) implies $f \circ \theta = f$ (P-a.s.).*

   (ii) *Let $f : \Omega \to \mathbb{R}$ be measurable. Then $(f - f \circ \theta) \in L^1(P)$ if, and only if, $(f - f \circ \theta)^+ \in L^1(P)$. In this case,*

$$E(f - f \circ \theta \mid \mathfrak{I}) = 0$$

P-a.s., where $\mathfrak{I} = \{F \in \mathscr{F} : P(\theta^{-1}F \triangle F) = 0\}$ is the σ-algebra of invariant sets. In particular, then $E(f - f \circ \theta) = 0$.

(*iii*) *If* $(f - f \circ \theta)^+ \in L^1(P)$ *then*

$$\lim_{n \to \infty} \frac{1}{n} f \circ \theta^n = 0 \qquad (B.2)$$

P-a.s. This applies, in particular, if $f \in L^1(P)$.

(*iv*) *For every* $\omega \in \Omega$, *for which* (B.2) *holds, also*

$$\lim_{n \to \infty} \frac{1}{n} \max_{0 \le k \le n} f \circ \theta^k(\omega) = 0.$$

PROOF (i) If $f \circ \theta < f$ with positive probability, then there exists $a \in \mathbb{R}$ with $P(f \circ \theta \le a < f) > 0$. However, $(f \le a) \subset (f \circ \theta \le a)$, hence $(f \le a) \cap (f \circ \theta \le a) = (f \le a)$, and so

$$(f \circ \theta \le a) = (f \le a) \cap (f \circ \theta \le a) \mathbin{\dot\cup} (f \circ \theta \le a < f) = (f \le a) \mathbin{\dot\cup} (f \circ \theta \le a < f)$$

($\dot\cup$ means disjoint union). We get

$$P(f \circ \theta \le a) = P(f \le a) + P(f \circ \theta \le a < f),$$

and thus $P(f \circ \theta \le a < f) = 0$ by θ-invariance of P.

(ii) One implication is immediate. To prove the other one suppose that $(f - f \circ \theta)^+$ is integrable. For $n \in \mathbb{N}$ put $f_n = (-n \vee f \wedge n) = (-n \vee (f \wedge n)) = ((-n \vee f) \wedge n)$, and note that $-n \vee (-f) \wedge n = -(-n \vee f \wedge n)$. We will show now that

$$(f_n - f_n \circ \theta)^+ \le (f - f \circ \theta)^+. \qquad (B.3)$$

Clearly this holds true if the left-hand side of (B.3) vanishes. In case it does not vanish,

$$-n \le f_n \circ \theta < f_n \le n,$$

hence $-n < f_n$, hence also $-n < f$, and so $f_n \le f$. On the other hand, $f_n \circ \theta < n$, and hence $f \circ \theta \le f_n \circ \theta$. Putting this together we obtain (B.3).

Applying the same argument to $-f$ we get

$$(f_n - f_n \circ \theta)^- \le (f - f \circ \theta)^-. \qquad (B.4)$$

Since $f_n \in L^1(P)$,

$$\int (f_n - f_n \circ \theta) \, dP = 0,$$

and thus

$$\int (f_n - f_n \circ \theta)^- \, dP = \int (f_n - f_n \circ \theta)^+ \, dP.$$

Now $(f_n - f_n \circ \theta)^-$ converges to $(f - f \circ f)^-$ P-a.s. with $n \to \infty$, so the Lemma of Fatou yields

$$
\begin{aligned}
\int (f - f \circ \theta)^- \, dP &= \int \liminf_{n \to \infty} (f_n - f_n \circ \theta)^- \, dP \\
&\leq \liminf_{n \to \infty} \int (f_n - f_n \circ \theta)^- \, dP \\
&= \liminf_{n \to \infty} \int (f_n - f_n \circ \theta)^+ \, dP \\
&\leq \int (f - f \circ \theta)^+ \, dP \\
&< \infty.
\end{aligned}
$$

Finally, note that $f_n \in L^1(P)$ implies

$$
E(f_n - f_n \circ \theta \mid \Im) = 0
$$

(since $E(f_n \circ \theta \mid \Im) = E(f_n \circ \theta \mid \theta^{-1}\Im) = E(f_n \mid \Im) \circ \theta = E(f_n \mid \Im)$). Now $(f_n - f_n \circ \theta)$ converges to $(f - f \circ \theta)$ P-a.s., and, by (B.3) and (B.4), $|f_n - f_n \circ \theta| = (f_n - f_n \circ \theta)^+ + (f_n - f_n \circ \theta)^- \leq (f - f \circ \theta)^+ + (f - f \circ \theta)^+ = |f - f \circ \theta|$. Therefore, by dominated convergence,

$$
0 = \lim_{n \to \infty} E(f_n - f_n \circ \theta \mid \Im) = E(f - f \circ \theta \mid \Im).
$$

(iii) First note that

$$
\frac{1}{n} f \circ \theta^n = \frac{1}{n} f - \frac{1}{n} \sum_{k=0}^{n-1} (f - f \circ \theta) \circ \theta^k.
$$

By (ii) we know that $(f - f \circ \theta) \in L^1(P)$. Therefore the individual ergodic theorem applies, yielding

$$
\lim_{n \to \infty} \frac{1}{n} f \circ \theta^n = E(f - f \circ \theta \mid \Im),
$$

which is equal to zero by (ii).

(iv) Suppose that, for a fixed $\omega \in \Omega$,

$$
\lim_{n \to \infty} \frac{1}{n} f(\theta^n \omega) = 0.
$$

For each $n \in \mathbb{N}$ there exists $k(n)$, $0 \leq k(n) \leq n$, with

$$
f(\theta^{k(n)} \omega) = \max_{0 \leq k \leq n} f(\theta^k \omega).
$$

Choose $\varepsilon > 0$, then there exists N_0 such that

$$\frac{1}{n}f(\theta^n\omega) < \varepsilon$$

for $n \geq N_0$. Put $F_n = \max_{0 \leq k \leq n} f(\theta^k\omega)$, and choose $N_1 > \varepsilon^{-1}F_{N_0}$, so that $n \geq N_1$ implies $\frac{1}{n}|F_{N_0}| < \varepsilon$. Estimating

$$\frac{1}{n}|F_n| = \frac{k(n)}{n}\frac{1}{k(n)}|F_n| \leq \frac{1}{k(n)}|f(\theta^{k(n)}\omega)| < \varepsilon$$

if $k(n) \geq N_0$, and

$$\frac{1}{n}|F_n| = \frac{1}{n}|F_{N_0}| < \varepsilon$$

if $k(n) \leq N_0$, we obtain $\frac{1}{n}|F_n| < \varepsilon$ for $n \geq \max\{N_0, N_1\}$. $\square$

If under the conditions of Proposition B.6 (i) the system (θ, P) is ergodic, then $f = \text{const } P$-a.s. If $(\theta_\alpha, P)_{\alpha \in A}$ is an ergodic family of measure preserving transformations, then $f \circ \theta_\alpha \leq f$ (P-a.s.) for every $\alpha \in A$ implies $f = \text{const}$ (P-a.s.).

Now let T be either $\mathbb{R}$, $\mathbb{R}^+$, $\mathbb{Z}$, or $\mathbb{N}$. For $t \in T$ denote by $[t] = \max\{n \in \mathbb{Z} : n \leq t\}$ the integer part of t, and denote by $\tau(t) = t - [t]$ its fractional part. Of course, this notation is superfluous for T discrete.

B.7 Corollary *Suppose that $(\vartheta_t)_{t \in T}$ is a flow on $(\Omega, \mathscr{F}, P)$ and $f : \Omega \to \mathbb{R}$ satisfies*

$$\sup_{0 < \tau \leq 1} (f \circ \vartheta_\tau - f)^+ \in L^1(P). \tag{B.5}$$

Then

$$\limsup_{t \to \infty} \frac{1}{t}(f \circ \vartheta_t) = 0$$

P-a.s.

PROOF Since

$$f \circ \vartheta_t = f \circ \vartheta_{\tau(t)} \circ \vartheta_{[t]} - f \circ \vartheta_{[t]} + f \circ \vartheta_{[t]} \leq \sup_{0 < \tau \leq 1} (f \circ \vartheta_\tau - f)^+ \circ \vartheta_{[t]} + f \circ \vartheta_{[t]},$$

the assertion follows from Proposition B.6 (iii) in view of the fact that (B.5) implies $(f \circ \vartheta_1 - f)^+ \in L^1(P)$. $\square$

B.8 Corollary *Suppose that $\theta : T \times \Omega \to \Omega$ is jointly measurable, where T is either $\mathbb{N}$, $\mathbb{Z}$, $\mathbb{R}^+$, or $\mathbb{R}$, such that θ_t is measure preserving for each $t \in T$. Let $f : \Omega \to \mathbb{R}$ be measurable, and suppose that $(f - f \circ \theta_\tau)^+ \in L^1(P)$ for some $\tau \in T$. Then P-a.s.*

$$\lim_{t \to \infty} \frac{1}{t} \int_t^{t+\tau} (f \circ \theta_s) \, ds = 0.$$

PROOF First note that

$$\int_t^{t+\tau} (f \circ \theta_s) \, ds = \int_0^\tau (f \circ \theta_s) \, ds + \int_0^t (f - f \circ \theta_\tau) \circ \theta_s \, ds.$$

By Proposition B.6 (ii), $(f - f \circ \theta_\tau)^+ \in L^1(P)$ implies $(f - f \circ \theta_\tau) \in L^1(P)$, and so, by the individual ergodic theorem,

$$\lim_{t \to \infty} \frac{1}{t} \int_t^{t+\tau} (f \circ \Theta_s) \, ds = \lim_{t \to \infty} \frac{1}{t} \int_0^t (f - f \circ \theta_\tau) \circ \theta_s \, ds = E(f - f \circ \theta_\tau \mid \mathfrak{I})$$

P-a.s., where $\mathfrak{I} = \{F \in \mathscr{F} : P(\theta_t^{-1} F \triangle F) = 0 \text{ for all } t \in T\}$ is the σ-algebra of invariant sets for the family $(\theta_t)_{t \in T}$. Since $\mathfrak{I}_\tau = \{F : P(\theta_\tau^{-1} F \triangle F) = 0\}$ satisfies $\mathfrak{I}_\tau \subset \mathfrak{I}$, Proposition B.6 (ii) implies $E(f - f \circ \theta_\tau \mid \mathfrak{I}) = 0$. $\square$

The assertion of the Corollary is almost immediate as soon as f is assumed to be integrable. Then the individual ergodic theorem implies that

$$\lim_{t \to \infty} \frac{1}{t} \int_0^t (f \circ \theta_s) \, ds$$

exists P-a.s., and coincides with

$$\lim_{t \to \infty} \frac{1}{t} \int_0^{t+\tau} (f \circ \theta_s) \, ds$$

P-a.s. The assertion follows by taking the difference.

Bibliography

[1] L. Arnold, *Random Dynamical Systems*, Springer, Berlin, 1998.

[2] J.-P. Aubin and H. Frankowska, *Set-Valued Analysis*, Birkhäuser, Boston, 1990.

[3] E. J. Balder, A general approach to lower semicontinuity and lower closure in optimal control theory, *SIAM J. Control Optimization* 22 (1984) 570–598.

[4] E. J. Balder, An extension of Prohorov's theorem for transition probabilities with applications to infinite-dimensional lower closure problems, *Rend. Circ. Mat. Palermo, II. Ser.* 34 (1985) 427–447.

[5] E. J. Balder, On Prohorov's theorem for transition probabilities, *Sémin. Anal. Convexe Montpellier II* Exposé 9 (1989) 9.1–9.11.

[6] H. Bauer, *Maß- und Wahrscheinlichkeitstheorie*, de Gruyter, New York, 1990.

[7] E. Behrends, *Maß- und Integrationstheorie*, Springer, Berlin, 1987.

[8] N. Bourbaki, *Eléments de Mathématique. Fascicule XXV, Livre Intégration*, Hermann & Cie, Paris, 1960.

[9] C. Castaing and M. Valadier, *Convex Analysis and Measurable Multifunctions*, Lecture Notes in Mathematics 580, Springer, Berlin, 1977.

[10] H. Crauel, Extremal exponents of random dynamical systems do not vanish, *J. Dyn. Differ. Equations* 2 (1990) 245–291.

[11] H. Crauel, Markov measures for random dynamical systems, *Stochastics Stochastics Rep.* 37 (1991) 153–173.

[12] H. Crauel, Non-Markovian invariant measures are hyperbolic, *Stochastic Processes Appl.* 45 (1993) 13–28.

[13] H. Crauel, Global random attractors are uniquely determined by attracting deterministic compact sets, *Ann. Mat. Pura Appl., IV. Ser.*, Vol. CLXXVI (1999) 57–72.

[14] H. Crauel, A. Debussche and F. Flandoli, Random attractors, *J. Dyn. Differ. Equations* 9 (1997) 307–341.

[15] H. Crauel and F. Flandoli, Attractors for random dynamical systems, *Probab. Theory Relat. Fields* 100 (1994) 365–393.

[16] H. Crauel and V. M. Gundlach (eds), *Stochastic Dynamics*, Springer, New York, 1999.

[17] D. Dawson and E. Perkins, *Historical Processes*, Memoirs of the American Mathematical Society 454, American Mathematical Society, Providence, 1991.

[18] C. Dellacherie and P. A. Meyer, *Probabilities and Potential*, North-Holland, Amsterdam, 1978.

[19] R. M. Dudley, *Real Analysis and Probability*, Wadsworth & Brooks/Cole, Pacific Grove, 1989.

[20] N. Dunford and J. T. Schwartz, *Linear Operators, Part I: General Theory*, Interscience, New York, 1958.

[21] A. M. Etheridge, *An Introduction to Superprocesses*, University Lecture Series 20, American Mathematical Society, Providence, 2000.

[22] S. N. Ethier and T. G. Kurtz, *Markov Processes, Characterization and Convergence*, Wiley, Chichester, 1986.

[23] F. Flandoli, *Regularity Theory and Stochastic Flows for Parabolic SPDEs*, Gordon and Breach Science Publishers, Amsterdam, 1995.

[24] P. Gänssler and W. Stute, *Wahrscheinlichkeitstheorie*, Springer, Berlin, 1977.

[25] P. R. Halmos, *Measure Theory*, Springer, New York, 1974.

[26] C. J. Himmelberg, Measurable relations, *Fund. Math.* 87 (1975) 53–72.

[27] O. Kallenberg, *Random Measures*, Akademie-Verlag, Berlin, 1975, and Academic Press, London, 1976.

[28] J. F. C. Kingman, The ergodic theory of subadditive stochastic processes, *J. Royal Statist. Soc. Ser. B* 30 (1968) 499–510.

[29] K. Kuratowski, *Topology, Vol. I*, Academic Press, New York, and PWN-Polish Scientific Publishers, Warszawa, 1966.

[30] G. Matheron, *Random Sets and Integral Geometry*, John Wiley & Sons, New York, 1975.

[31] J. C. Oxtoby, Ergodic sets, *Bull. Amer. Math. Soc.* 58 (1952) 116–136.

[32] K. R. Parthasarathy, *Probability Measures on Metric Spaces*, Academic Press, New York and London, 1967.

[33] Yu. V. Prohorov, Convergence of random processes and limit theorems in probability theory, *Theory Probab. Appl.* 1 (1956) 157–214.

[34] B. v. Querenburg, *Mengentheoretische Topologie*, Zweite, neubearbeitete und erweiterte Auflage, Springer, Berlin, 1979.

[35] W. Rudin, *Functional Analysis*, McGraw-Hill, New York, 1973.

[36] R. Temam, *Infinite-Dimensional Dynamical Systems in Mechanics and Physics*, Second Edition, Springer, New York, 1997, 1988.

[37] M. Valadier, Young measures, pp. 152–188 in *Methods of Nonconvex Analysis, Varenna 1989*; A. Cellina (ed.), Lecture Notes in Mathematics 1446, Springer, Berlin, 1990.

[38] P. Walters, *An Introduction to Ergodic Theory*, Springer, New York, 1982.

[39] D. Williams, *Diffusions, Markov Processes, and Martingales*, Wiley, Chichester, 1979.

Index